BE SAT READY! CROSSWORDS FOR SAT

Hard Crossword Puzzle Books (with **50** drills)

PUZZLE THERAPIST

CROSSWORD | SUDOKU | KIDS & ADULTS

CONTENT

PUZZLE 1

ACROSS

1. Fill in the blank with this word: "___ fu"

5. Wood-cutting tool

9. Hypothetical physics particle

14. What's ___ for me?'

15. Pas ___ (gentle ballet step)

16. Maui neighbor

17. 1989 Al Pacino movie

19. Woodworking fasteners

20. Fill in the blank with this word: "Amniotic ___"

21. Now ___ me down to sleep'

22. The Godfather' co-star

23. Star QB for the 1980s-'90s Bengals

25. Noted Parthenon sculptor

28. Lip-___

29. Fill in the blank with this word: "___ tai"

30. Naut. law enforcers

31. Jack-in-the-box needs

34. A Thing ___' (Beach Boys song)

35. Vacation spot

38. Fill in the blank with this word: "Correo ___"

39. Lennon's in-laws

40. Wellness grp.

41. The annus in Dryden's "Annus Mirabilis"

43. Fill in the blank with this word: ""Deutschland ___ Alles""

45. They liaise with the FBI

46. Successor to the Cutlass

50. Ship out

52. You can't focus when this is on

54. Verb type: Abbr.

55. Unscramble this word: neam

56. Therefore

57. Union part

59. Early Lerner and Loewe hit

61. Thereby hangs ___'

62. Actress Skye

63. Start of a string of 13 Popes

64. Old Italian coin

65. One succumbing to 6-Down

66. George Harrison's "___ It a Pity"

DOWN

1. Things wrapped in foil

2. Restless

3. Pellagra preventer

4. Fill in the blank with this word: ""Little ___, you're really lookin' fine" (1964 lyric)"

5. Verdi's "Un ___ in Maschera"

6. Fill in the blank with this word: "___-Kettering Institute"

7. Woody's role in "Annie Hall"

8. Just a ___ bit

9. Russia's ___ Republic

10. Fill in the blank with this word: "___ 2.0, Bill Gates's house"

11. Together

12. Kellogg's Cracklin' ___ Bran

13. Sue Grafton's '___ for Noose'

18. The English translation for the french word: pas catholique

22. Longtime Comiskey Park team, informally

24. Fill in the blank with this word: "Caladryl: itch :: Bengay : ___"

25. Fill in the blank with this word: "Control ___"

26. Top: Prefix

27. Three-stripers: Abbr.

29. Some autos, for short

32. Mollycoddle

33. Ibsen's Gabler

35. Repeated cry at a beer blast

36. Early pulpit

37. Using high-flown language

39. The Van Eyck bros. weren't the 1st to use these paints, though sometimes creditied with it; they date back much earlier

42. Woman of letters?

44. Humorous illustrator ___ Searle

47. They repeat whatever you say

48. an omnivorous nocturnal mammal native to North America and Central America

49. Urging at a birthday party

51. Pope with a Nov. 10 feast day

52. Hanover's river

53. Zealous

55. Ancient Greek sculptor of athletes

57. Velázquez's '___ Meninas'

58. Fill in the blank with this word: "Cardio : heart :: ___ : ear"

59. Twice, in music

60. Year in Vigilius's papacy

PUZZLE 2

1	2	3	4		5	6	7	8		9	10	11	12	13
14					15					16				
17					18					19				
20				21				22			23			
				24				25			26			
27	28	29	30			31	32				33	34	35	36
37				38		39				40				
41					42			43						
44				45			46							
47				48		49		50						
		51	52				53	54						
55	56	57		58		59				60	61	62	63	
64		65			66				67				
68					69				70				
71					72				73				

ACROSS

1. Swimming great Diana

5. City in Judah

9. To say in Spanish?

14. The English translation for the french word: gala

15. Fill in the blank with this word: "____ one"

16. How a trucker might go up a hill

17. Fill in the blank with this word: ""____ my flesh of brass?": Job"

18. Jacquerie

19. Pear variety

20. Appetizer that diverts attention from the main menu?

23. Paul Scott series "The ____ Quartet"

24. War on Poverty agcy.

25. Fill in the blank with this word: ""____ help you?""

27. Unscramble this word: htleo

31. Songwriter Jacques

33. Some shoes, for short

37. "Savvy?"

39. Tchaikovsky's Symphony No. 5 ____

minor

40. Times past

41. Santa ___

44. Touched the tarmac

45. New York ___

46. Pesters

47. Ovid's family name

48. Fill in the blank with this word: ""The even mead, that ___ brought sweetly forth ...": "Henry V""

50. B and O figures: Abbr.

51. What knows the drill, for short?

53. Bee: Prefix

55. Fill in the blank with this word: "___-Rhin (Strasbourg's department)"

58. Where weapons are forbidden

64. Unoriginal

66. Unscramble this word: stso

67. Wild Indonesian bovine

68. Relating to form

69. Radio host Don

70. Millennium divs.

71. Snares

72. Trans-Siberian Railway hub

73. Fill in the blank with this word: ""And ___ thou

slain the Jabberwock?""

DOWN

1. Haing S. ___ (Oscar winner for 'The Killing Fields')

2. Responsive, as a ship

3. Put ___ on it!'

4. Word repeated in "Now ___ away! ___ away! ___ away ...!"

5. Harden

6. Sure thing!'

7. The Bell of ___' (Longfellow poem)

8. Primary computer list

9. Scatterbrained

10. Photography abbr.

11. Sharp

12. Whit

13. Medieval romance tale

21. Ancient Greek tongue: Var.

22. PbS

26. Clarifying words

27. Adm. Rickover of the 40-Across

28. Alley Oop's mate

29. Gulf of ___ (arm of the Mediterranean)

30. Writer's aid

32. Unrest

34. You can hunt down 2 of the 10 brightest stars in the night sky in this constellation

35. Echo producer

36. Exceptional rating

38. The English translation for the french word: gamËte

42. Tybalt kills him

43. Kind of sketch

49. St. Paul's birthplace

52. Starting Harvard law students

54. House detective's item

55. Rec rm. locale, often

56. Suffix with origin

57. Thompson of "Family"

59. With 6-Down, 1994 Olympic gold medalist in downhill skiing

60. Fill in the blank with this word: ""Scrubs" co-star ___ Braff"

61. Fill in the blank with this word: "___-Day"

62. Votes in Versailles

63. White House's ___ Room

65. Unscramble this word: alp

PUZZLE 3

1	2	3	4	5	6	7	8		9	10	11	12	13	14
15									16					
17									18					
		19				20								
21	22	23					24			25	26	27	28	
29				30	31	32				33				
34			35					36						
	37	38				39	40							
41					42					43		44		
45				46	47				48					
49			50				51							
		52			53	54	55							
56	57	58	59			60				61	62	63		
64						65								
66						67								

ACROSS

1. Like some specially-prepared foods

9. 1977 James Brolin thriller with the tagline "What EVIL drives ..."

15. Enthusiastic response

16. Person of high position

17. a substance added to another to make it less hard

18. Actress Massey et al.

19. Supplication

21. Of ___ (somewhat)

24. New Jersey's ___ Air Force Base

29. Throat stuff

30. Cover-up

33. Keglers' places

34. Rock's ___ Speedwagon

35. Weekend Today' anchor Lester

36. These gastropods are sometimes fed aromatic herbs to give them a special savor

37. Pop artist whose name is an anagram of 20-Across

41. Film director Pier ___ Pasolini

42. Detestation

43. Misery

45. Start of Massachusetts' motto

46. Fill in the blank with this word: ""It is equally an error to ___ all men or no man": Seneca"

48. Common contraction

49. Prankster's cry

51. Kentucky's ___ College

52. "Carmen" highlight

56. Union general Henry Warner ___

60. Made a comeback?

64. Guy Lombardo hit of 1937 or Jimmy Dorsey hit of 1957

65. Skip a dinner date

66. Tangles

67. Prepare in advance of

DOWN

1. Trash

2. Words said at an altar

3. Jolly old ___ (Santa)

4. Until the due date

5. Fill in the blank with this word: "Blessed ___"

6. Unscramble this word: nett

7. Route for Ben-Hur

8. Rotten to the ___

9. Spasm

10. Trinity member

11. Mancinelli's "___ e Leandro"

12. Whse. unit

13. Fill in the blank with this word: "___ moment"

14. Trains: Abbr.

20. Pro-___ (some tourneys)

21. Fill in the blank with this word: "0% ___"

22. Whole ___

23. Wine whose name means "fragrant"

25. Wrist-elbow connector

26. Well-suited?

27. Tenth-anniversary DVD, e.g.

28. What you might hear halting speech in, for short

30. Rock's ___ Jovi

31. Corduroy ridge

32. Right back ___!'

35. Water

36. Blood bank supply

38. Frankie Laine's "___ Her Go"

39. Vietnam War-era org.

40. Word with boss or bull

41. Shar-___ (wrinkly dog)

44. Substance checker:

Abbr.

46. View from Windsor Castle

47. Music category

48. Like the return of swallows to Capistrano

50. Boorish sort

51. Vast

53. War maker

54. Fill in the blank with this word: "___ as a pin"

55. Fill in the blanks with these two words: ""___, Brute?""

56. Worrying sound to a balloonist

57. Fill in the blank with this word: "Cambodia's ___ Nol"

58. Fill in the blank with this word: "___ pro nobis"

59. Fill in the blank with this word: "Elevator ___"

61. Fill in the blank with this word: "___ a kind (pair)"

62. Rock's Brian ___

63. Wagner's "___ fliegende Hollander"

PUZZLE 4

1	2	3	4		5	6	7	8		9	10	11	12	13
14					15					16				
17				18					19					
20				21						22				
			23			24			25		26			
27	28	29			30			31		32				
33					34	35								
36					37					38	39	40	41	
		42	43	44			45	46						
	47	48				49								
50			51		52	53		54						
55		56		57		58				59	60	61		
62			63					64						
65				66				67						
68				69				70						

ACROSS

1. Old record label

5. One who follows the news

9. Fill in the blank with this word: ""May ___ your order?""

14. Ostentatious display

15. Writer of sweet words?

16. Golf's ___- Ryder Open

17. Like the thief at 17-Across

20. Winding road shape

21. Modify anew

22. Fill in the blank with this word: "Auvers-sur-___, last home of Vincent van Gogh"

23. Fill in the blank with this word: ""___ Woman" (1972 #1 song)"

24. Wrinkly-faced dogs

26. Fill in the blank with this word: "China's Sun Yat-___"

27. Set up

31. Getty Center architect Richard

33. Go ballistic

36. Summer coolers

37. Word with bum or bunny

38. The Isle of Man's Port ___

42. Northern flier / Mixer maker / Put on the line

47. Fill in the blank with this word: ""___ Meets Godzilla" (classic 1969 cartoon)"

49. What shall be first ... or words that can precede 17-, 23-, 52- and 60-Across

50. U.N. chief ___ Ki-moon

51. Top: Prefix

54. U.S.N.A. grad

55. Yearbook sect.

57. Set upon

59. Yemen-to-Zimbabwe dir.

62. Might-have-been Midwest team

65. The English translation for the french word: saint

66. Fill in the blank with this word: "Comic strip "___ & Janis""

67. Word: Suffix

68. The English translation for the french word: humilier

69. Fill in the blank with this word: "Coral ___"

70. Loch ___ monster

DOWN

1. Oxford bottom

2. Wooden beams laid down to secure the rails of a railroad

3. Warehouse supply: Abbr.

4. Fever reading, maybe

5. Telephone riggers

6. Fill in the blank with this word: "___ signum (here is the proof)"

7. The English translation for the french word: attente

8. Fill in the blank with this word: "Do-___"

9. Words said at an altar

10. Axis leader

11. Get ___ (be rewarded at work)

12. Fill in the blank with this word: "___ roll"

13. Popular 1990's sitcom

18. Wordsworth's Muse

19. Cheat on

23. Winter river obstruction

25. Yellow ___

27. Fill in the blank with this word: "England's Isle of ___"

28. Cartoonist Chast

29. Yacht's dir.

30. Fam. tree member

32. Worthy principles

34. ___ punk (hybrid music genre)

35. Touch

39. Vitamin stat.

40. Fill in the blank with this word: "___ agent"

41. WSJ competitor

43. Hall-of-Fame basketball coach Hank

44. Perfume maker Nina ___

45. Leading

46. Tooth: Prefix

47. Zombies might be on it

48. Land south of Hadrian's Wall

50. Fill in the blank with this word: "___ nova"

52. Buddy of 1940's baseball

53. Samuel Lover's "Rory ___"

56. Princes, e.g.

58. Writer ___ Stanley Gardner

59. Within reason

60. Stallone and others

61. Circus reactions

63. Sport-___ (vehicle)

64. Fill in the blank with this word: "___ compos mentis"

PUZZLE 5

1	2	3	4	5		6	7	8	9		10	11	12	13
14						15					16			
17				18					19					
20					21				22					
		23	24				25							
26	27	28		29		30	31					32	33	34
35				36							37			
38			39			40			41					
42					43			44			45			
46			47	48							49			
		50						51		52				
53	54	55				56	57			58		59	60	61
62					63				64					
65				66					67					
68				69					70					

ACROSS

1. Fill in the blank with this word: "___ Picchu, Peru"

6. Toiletry item

10. Michael of 'Juno'

14. Fill in the blank with this word: "___ Island National Monument"

15. Sheik ___ Abdel Rahman

16. Zip

17. Fill in the blank with this word: """I noticed you use the ___ ___ often than the tarnished one""

20. Fill in the blank with this word: """___ be in England...""

21. Fill in the blank with this word: "___ show"

22. Idiot box

23. Rum ___ Tugger (

25. Fill in the blank with this word: "___ Southwest Grill (restaurant chain)"

26. ___ port

29. They won't wait, in a phrase

35. Sports org.

36. Sniggled

37. Poisonous Asian plant

38. Retail clothing giant ... or a description of 17- and 54-Across and 10- and 24-Down?

40. Org. with an annual televised awards ceremony

41. Exceptional rating

42. Ireland's ___ Islands

43. Tops

45. Obama's signature health law, for short

46. Double-check, say

49. Fill in the blank with this word: ""Tais-___!" (French "Shut up!")"

50. Waterfall sound

51. White House advisory grp.

53. Wrinkle preventer, of sorts

56. QuÈbec's ___ d'OrlÈans

58. Wall St. deals

62. They're dishwasher-safe

65. Part of QE2: Abbr

66. Indian of the Sacramento River valley

67. Sufficiently old

68. Word game component, sometimes

69. Too much ink

70. Fill in the blank with this word: "___-O-Matic (maker of sports games)"

DOWN

1. The Bible Tells ___'

2. Xanadu river

3. When fibrogen is converted to fibrin by thrombin, blood does this

4. Fill in the blank with this word: "___-miss"

5. Southeastern Conf. team

6. Unscramble this word: stso

7. Fill in the blank with this word: ""I know not why I ___ sad": Shak."

8. Unscramble this word: alp

9. New York's ___ Falls

10. Some metallophones

11. Your highness?: Abbr.

12. Tear down, in England

13. Handle: Fr.

18. Jazzman Blake

19. Ticked (off)

24. Beginning on: 2 wds.

25. Puccini's "___ Butterfly"

26. Fix, as a printer's feeder

27. Pelvis connectors

28. P.M. between Netanyahu and Sharon

30. Celtics head coach, 1995-97

31. Famed statement by 67-Down

32. Where ___

33. The English translation for the french word: disco

34. Writing by Montaigne

39. Rev up

41. Workers need them: Abbr.

44. German indefinite article

47. Filmmakers: Joel & Ethan

48. Kind of special

52. Split

53. Operator's accessory

54. Mythical king of the Huns

55. Tenpenny ___

56. This __ laughing matter!'

57. Would-be J.D.'s hurdle

59. You have the right to do this regarding arms, but your arms will be this without sleeves

60. Fill in the blank with this word: "2000 Olympic hurdles gold medalist ___ Shishigina"

61. Fixed at an acute angle

63. Suffix with ether

64. Talking-___ (scoldings)

PUZZLE 6

ACROSS

1. Freud contemporary

6. Steepness

11. Cries of pain

14. Fill in the blank with this word: "___ Oro"

15. World's smallest island nation

16. Panama, e.g.

17. someone who works metal (especially by hammering it when it is hot and malleable)

19. Fill in the blank with this word: "Electric ___"

20. Fill in the blank with this word: ""Humanum ___ errare""

21. Fill in the blank with this word: ""Je vous ___""

22. Volunteer

24. "Prepare to do a spoof on airports"

28. Fill in the blank with this word: ""Beauty ___ the eye Ö""

29. Fill in the blank with this word: "___ II (razor brand)"

30. The Kennedys, e.g.

33. Suffix with Meso- or Paleo-

34. Year of Bush's swearing-in

37. Region with the highest concentration of national parks in the U.S.

42. On the ___

43. Michigan's ___ College

44. "Excuse me Ö"

45. Fill in the blank with

this word: "Cup ___ (hot drink, informally)"

46. Watch face

49. Rolling Stones hit of 1968

56. Swindler

57. Openness

58. Mao's mil. force

59. Sir ___ McKellen (Gandalf portrayer)

60. A pharaoh vis-

64. Moli√°re's 'Le M√©decin Malgr√© ___'

65. Fill in the blank with this word: "___ vie"

66. What is the capital of this country - Belarus

67. Slippery ___

68. Fill in the blank with this word: "___ good example (shows the proper way)"

69. Tallinn natives

DOWN

1. Ordnance supplier

2. This, in Th

3. "Little" girl of old comics

4. Soprano Christiane ___-Pierre

5. Some races

6. Part of a sundial that casts a shadow

7. With more to be done?

8. You'll find this bird in Milwaukee

9. Rap's Dr. ___

10. 95501

11. New Mexico county

12. Fill in the blank with this word: "By ___ (via)"

13. Fill in the blank with this word: "Continental ___"

18. Provider of a hot spot at a coffee shop?

23. Toshiba competitor

25. Purcell's "___ and Aeneas"

26. 'Vette option

27. Seed covering

30. Basketball coach Jones and others

31. Fill in the blank with this word: "___ Hamoed (Passover period)"

32. Round-the-world traveler Nellie

33. Taxonomy suffix

34. Yawn

35. Writer Rita ___ Brown

36. Elementary suffix

38. Confirms

39. Fill in the blank with this word: "Chrysler Building architect William Van ___"

40. Key with two sharps:

Abbr.

41. Fill in the blank with this word: "Albee's "Three ___ Women""

45. To his good friends thus wide I'll ___ my arms': 'Hamlet'

46. Where a lot of fed. govt. workers live

47. Venerated image: Var.

48. Structure that's roughly a triangular prism

49. Scrabble piece not found in the Italian edition

50. Trailing business?

51. The English translation for the french word: blanche

52. Schools for cadets: Abbr.

53. Hoist ___ (enjoy the pub)

54. What water in a pail may do

55. Tom who played Forrest Gump

61. Scottish refusal

62. Studio shout

63. Word repeated in Emily Dickinson's "___ so much joy! ___ so much joy!"

PUZZLE 7

1	2	3	4		5	6	7	8	9		10	11	12	13
14					15						16			
17					18						19			
20			21							22				
			23				24	25						
26	27	28	29			30	31							
32					33		34				35	36	37	38
39					40	41					42			
43					44				45	46				
			47	48				49		50				
51	52	53						54	55					
56					57	58					59	60	61	62
63				64						65				
66				67						68				
69				70						71				

ACROSS

1. Plasm prefix

5. Variety has long used this word for a box office hit

10. Nonsense

14. The English translation for the french word: naÔf

15. Two strikes?

16. Tiers ___ (French commons)

17. Tournament passes

18. Part of a metropolitan region

19. Gillette ___ Plus

20. tough elastic tissue

22. Radical

23. Inc. cousin

24. Flower parts

26. Unite

30. Winner of all four grand slam titles

32. Fill in the blank with this word: "___ del Fuego"

34. Cries of pain

35. Sgts., e.g.

39. Fit of shivering, in dialect

40. Early statistical software

42. Zipped

43. Zaire's Mobutu ___ Seko

44. Yahoo! competitor

45. Phrase of irresolution

47. Shipping weight

50. Fill in the blank with this word: "Correo ___"

51. Sewing groups

54. Son of, in Arabic names

56. Fill in the blank with this word: "Allegro ___ (very fast)"

57. Viniculturist's sampling tube

63. Fill in the blank with this word: "Dragon's ___ (early video game)"

64. Holy, to Horace

65. Mae West's '___ Angel'

66. Volunteer org. launched in 1980

67. When lunch ends, maybe

68. Mineral residue

69. Word with code or road

70. Loosens (up)

71. With 52-Across, what angels pray for

DOWN

1. You might take investing tips from this network's "On the Money" or "The Call"; Jon Stewart probably doesn't

2. Fill in the blank with this word: ""Divine Secrets of the ___ Sisterhood""

3. Top-___ (leading)

4. Fill in the blank with this word: ""The Bells ___ Mary's""

5. Lymphocyte found in marrow

6. State of southern Mexico

7. Lively '60s dance

8. Raccoon's hands

9. White House fiscal grp.

10. With 59-Across, indication of caring

11. Holy Roman emperor, 962-73

12. Pope John Paul II's real first name

13. The Louvre's Salles des ___

21. The Sopranos' actor Robert

22. Worrying sound to a balloonist

25. Japanese immigrant

26. Chairmen often call them: Abbr.

27. Yeats's land

28. Defendant at law

29. Work necessities, for some

31. With 70-Down, do much (for)

33. It's ___!' ('We'll go out together!')

36. Rope fiber

37. Writer Sarah ___ Jewett

38. Zaire's Mobuto Sese ___

41. Neighbor of South Africa

46. Fill in the blank with this word: ""___ pis!" ("Too bad!," in France)"

48. Ransom ___ Olds

49. Prime-time time

51. Serenity, in Seville

52. Writer Asimov

53. Where to sign a credit card, e.g.

55. Earthen embankments

58. Writer of sweet words?

59. Bar sounds

60. The English translation for the french word: imam

61. Requests for developers: Abbr.

62. Below-ground sanctuary

64. You reap what you ___'

PUZZLE 8

1	2	3	4		5	6	7	8		9	10	11	12	13
14					15					16				
17					18			19						
20			21			22								
23				24	25						26	27	28	
29				30				31		32				
		33				34	35			36				
37	38	39						40	41					
42					43									
44			45	46			47				48	49	50	
51			52		53	54				55				
		56							57					
58	59	60					61	62		63				
64					65					66				
67					68					69				

ACROSS

1. Assns. and orgs.

5. The ___ Nugget, Alaska's oldest newspaper

9. Auto option

14. City on the Gulf of Aqaba

15. Fill in the blank with this word: "Author ___ S. Connell"

16. Norman Vincent ___

17. Fill in the blank with this word: "___ tide"

18. Like Buddy Holly's glasses

20. Shame ___!'

22. Thirst (for)

23. Russian's neighbor

26. Fill in the blank with this word: "___ Cayes, Haiti"

29. Yes, ___'

30. Every, to a pharmacist

31. Optimally

33. Will words

36. Battle of ___ (1943 U.S./Japanese conflict)

37. Source of bolts

42. Smear with wax, old-style

43. Longtime radio advice-giver

44. With no pence on hand, you might ask the curry restaurant, "Will you take" a personal one of these?

47. TV schedule abbr.

48. White House inits.

51. Sue Grafton's '___ for Evidence'

52. "Shake a leg!"

56. Noted diamond family

57. Like some glasses, in a phrase

58. Parachute pack attachment

63. Zebras

64. Venae ___ (major blood vessels)

65. Verified, in a way

66. Fill in the blank with this word: ""___ can't be!""

67. Many ___ (a long time)

68. Yappy dog, briefly

69. Workers need them: Abbr.

DOWN

1. Racing jibs

2. Wagner opera based on a 14th-century Italian patriot

3. Ladies' man

4. Flavor

5. Old Testament book: Abbr.

6. Omne vivum ex ___ (all life [is] from eggs: Lat.)

7. Pot

8. Nine: Prefix

9. Unscramble this word: tpiisr

10. Staff differently

11. When "S.N.L." wraps in N.Y.C.

12. Grand ___ Opry

13. Spoon-___

19. Fill in the blank with this word: "___ temperature (was feverish)"

21. Nut holder

24. Western Hemisphere abbr.

25. Grants-___

26. Mother of Apollo

27. Old Testament book

28. Wine additive

32. Turkish ___

33. Miss Congeniality she's not

34. TV/___

35. Mother ___

37. Fill in the blank with this word: "___ signum (here is the proof)"

38. Where Samson wielded a jawbone, in Judges

39. Fill in the blank with this word: ""So ___ to you, Fuzzy-Wuzzy": Kipling"

40. Truck treatment

41. Kind of infection, for short

45. Hags, e.g.

46. Workplace fairness agcy

48. Lumberjack contests

49. Breakfast bread

50. Outlawed blasts

53. Turban & this flower name share the same Turkish roots

54. Up on deck

55. They're raised on farms

56. Mussorgsky's 'Pictures ___ Exhibition'

58. Inits. on old typewriters

59. Way of the East

60. Cosmetics giant

61. Educational cable network

62. Dutch city

PUZZLE 9

1	2	3	4		5	6	7	8		9	10	11	12	13
14					15					16				
17					18					19				
20				21				22				23		
			24						25		26			
27	28	29		30					31			32	33	34
35			36			37	38	39			40			
41				42						43				
44				45						46				
47				48			49	50			51			
			52			53					54			
55	56	57		58		59					60	61	62	
63			64			65				66				
67					68				69					
70					71				72					

ACROSS

1. Watchdog org.?

5. Troubles

9. Fill in the blank with this word: ""... ___ the queen of England!""

14. Fill in the blank with this word: "Albee's "Three ___ Women""

15. Fill in the blank with this word: "___ monde"

16. Witherspoon of "Vanity Fair"

17. Jewel

18. Fill in the blank with this word: "___ in a blue moon"

19. Serenity, in Seville

20. Clear orders

23. Unrealized 60's Boeing project

24. Joe Jackson's "___ Really Going Out With Him?"

25. Fill in the blank with this word: "Black-throated ___"

27. Union ___: Abbr.

30. USN enlistee

31. Set, as a gem

35. U.S. Open champ, 1985-87

37. Obama adviser Emanuel

40. You might stick a knife in it

41. Game settings

44. Jazz singer ___ James

45. Uncle ___

46. #2's

47. TV's "___ and Greg"

49. Woe ___' (popular grammar book)

51. Pitcher Robb ___

52. It was split in 1948: Abbr.

53. Ivy Leaguer's home

55. Wing it?

58. "Do tell!"

63. Mysterious

65. You might give a speech by this

66. Performance halls

67. When you "make" this, you go with speed

68. In ___ (worked up)

69. Dogpatch possessive

70. Some sneakers

71. Scottish blackbird

72. Kipling's "___ we forget!"

DOWN

1. Naut. direction

2. Au ___

3. Surfeit

4. Food preserver?

5. What Richard III offered "my kingdom" for

6. 2004 Olympics swimming star

7. Saint ___ (Florida county)

8. World War II weapon

9. It's measured in radians

10. Fill in the blank with this word: "___ Today (teachers' monthly)"

11. Pol. convention attendees

12. Things to believe in

13. Vegetarian's no-no

21. Shelters

22. Twos in the news

26. Actress Birch of "American Beauty"

27. Set upon a slope, say

28. Tithing portion

29. Santa ___, California track

32. Singing chipmunk

33. The Beatles' "You Won't ___"

34. Rose-red dye

36. German currency, informally

38. Son of Prince Valiant

39. one who hesitates (usually out of fear)

42. Steel braces with right-angle bends

43. Fill in the blank with this word: "Dr. ___ Hahn of 'Grey's Anatomy'"

48. Shimmery fabrics

50. "Fiddler on the Roof" setting

53. Fill in the blank with this word: "'___ you're satisfied now!'"

54. Make ___ of (embarrass)

55. Remarked

56. Where to wear a muumuu

57. Quaint affirmative

59. Way around London, once

60. Together, to Toscanini

61. Football linemen

62. Fill in the blank with this word: "'___ pis!' ("Too bad!," in France)"

64. Fill in the blank with this word: "___ Aquarids (May meteor shower)"

PUZZLE 10

ACROSS

1. Writer's woe

6. They, in Italy

10. Sound of a leak

14. Lima's land, to the French

15. Call in the game Battleship

16. Fit ___ (be perfect on)

17. I Kissed ___' (Katy Perry hit)

18. Wear down

19. Sen. McCain's state: Abbr.

20. Working at an auto plant David Smith learned the techniques he used sculpting this ‚Äústainless‚Äù metal

21. Zenith products

22. Fill in the blank with this word: "___ temperature (was feverish)"

23. Sports org. that publishes DEUCE magazine

25. Seacrest's 'American Top 40' predecessor

27. Towering desert plants: Var.

31. Sings hallelujah to

35. Wrath

36. Hezbollah stronghold ___ Valley

38. Part of a pay-as-you-go plan?

39. The X in this holiday spelling comes from the Greek letter chi & also represents the cross

41. Standing Bear's tribe

43. Unscramble this word: lsle

44. University of Maryland, informally

46. Facing the pitcher: 2 wds.

48. Fill in the blank with this word: "___ el Amarna, Egypt"

49. Work of 1599

51. Kind of order to a broker

53. Unlikely to reconsider

55. Sue Grafton's '___ for Ricochet'

56. New York's Carnegie ___

59. Openness

61. Suggest subtly

65. Word said just before opening the eyes

66. Unpopular spots

67. Once-___ (quick appraisals)

68. Cut down on

69. Scottish rejections

70. Way cool

71. Town line sign abbr.

72. Sitting spot

73. Spanish beings

DOWN

1. Their days are numbered

2. Mil. unit below a division

3. Singer India.___

4. Dr. in an H. G. Wells novel

5. Soda can feature

6. Ron Howard media satire

7. Start of a playground chant

8. Missionary Junipero ___

9. Really here

10. Some Beverly Hills tourist purchases

11. Short-billed rail

12. 1940's-50's All-Star Johnny

13. Anthem part

24. Warm-up

26. Phone no. add-on

27. With 52-Down, intuition

28. Napol

29. Two-time Tony winner George

30. Relenting assent

32. Stay stationary with bow to windward

33. Bell sounds

34. Unloads

37. Fill in the blank with this word: "___ no."

40. Twisted in a bad way

42. Get an ___ effort

45. One of the Chaplins

47. Undistinguished imitator

50. This adjective for an empty lot or uninhabited house also refers to an estate that no heir has claimed

52. Square, in 1950s slang, indicated visually by a two-hand gesture

54. Hong Kong neighbor: Var.

56. Truth or ___ (slumber party game)

57. Scottish uncles

58. Univ. worker

60. Letter before shin

62. The English translation for the french word: tÈtine

63. Vissi d'___' (Puccini aria)

64. General ___ chicken (Chinese menu item)

PUZZLE 11

1	2	3	4	5		6	7	8	9		10	11	12	13
14						15					16			
17						18					19			
20				21						22				
			23					24	25					
26	27	28	29		30				31					
32				33			34		35		36	37	38
39			40				41	42					
43					44	45				46			
		47		48		49				50			
51	52	53				54			55					
56					57	58					59	60	61	62
63				64				65					
66				67				68					
69				70				71					

ACROSS

1. Philippine banana plant

6. Famous brother

10. European freshwater fish

14. Trivial Pursuit edition

15. Wolfe, the sleuth

16. Penultimate word in a fairy tale

17. 'Bellefleur" author

18. Cream was one

19. Expurgate, editorially

20. A detailed plan

22. Vena --- (vessel to the heart)

23. Native American tent

24. Nonpointed end

26. Hit alternative

30. Networked computers, for short

31. Pencil stump

32. Agency concerned with civil aviation

33. Kim Jong-il's place

35. Type of iron girder

39. Undone or leave†out

41. Pay for, as a project

43. Father of Jacob

44. Something fishy?

46. Word with "movie" or "party"

47. Pythagorean P

49. Romantic or Victorian, e.g.

50. Where starter

51. Berry the size of a hen egg

54. Kind of blocker

56. Endangered buffalo

57. China or silverware

63. Common subject?

64. Dangerous marine creature

65. Subject of a house inspection test

66. Between gigs

67. Fruit with a wrinkled rind

68. Not under one's breath

69. Holding place

70. Hammered at a slant

71. Puppies' cries

DOWN

1. Slack-jawed

2. Gal's sweetheart

3. Contrary one

4. Handed a line

5. Balance sheet plus

6. Italian meal starter, perhaps

7. Italian sculptor

8. Albany canal

9. Undo, in a way

10. Head of purplish-red leaves

11. Eye layers

12. Dig deeply

13. Bleak, in verse

21. Lavender flower

25. Remains to be seen?

26. Abbreviated version

27. Tries to reduce swelling, in a way

28. Long story

29. Electrical device

34. Raised above?

36. Organic compound

37. Blackjack components

38. Word with pittance

40. Canyon sound

42. Hot under the collar

45. Humiliating failure

48. Decline an invitation

51. Frenzied

52. Battery terminal

53. '___ be sorry!"

55. Arrangement

58. Thus

59. Low tract

60. Fanzine focus

61. ___ d'etat

62. Wraps up

PUZZLE 12

1	2	3	4		5	6	7	8		9	10	11	12	13
14					15					16				
17					18					19				
20				21					22					
	23						24					25	26	27
			28			29					30			
31	32	33		34							35			
36			37					38	39	40				
41							42					43		
44				45	46	47					48			
49				50					51			52	53	
		54					55							56
57	58	59				60					61			
62						63					64			
65						66					67			

ACROSS

1. Obtained from urine

5. Double-reed woodwind

9. Elongated fruit (Var.)

14. Sacred Hindu writing

15. Prison sentences

16. Cover story?

17. Make uniform

18. Intl. commerce pact

19. Skirt fold

20. Australia's national day

23. Distinctive spirit of a culture

24. Relating to the ear

25. 'That's all ___ wrote"

28. Protects the body from foreign substances

31. Pint-size

34. Is not well

35. Of the first water

36. Supplement

38. Not likely to get by the censor

41. Scurries

42. Civil aviation

43. Golf standard

44. From time to time

49. Baseball great Mel

50. Current jumps, e.g.

51. Loamy deposit

54. Friend or competitor

57. Grape seeds

60. Invited a perjury charge

61. Lion's pride

62. Rainwater pipe

63. Strongly suggest

64. Aquarium dweller

65. Flower from the violet family

66. Famous septet

67. Where worms may be served

DOWN

1. Part of the eye

2. Las Vegas show, perhaps

3. Cato's clarification

4. Upper and lower eyelids meet

5. Very pleasurable

6. False god

7. Alternatives

8. Beverly Hills home, typically

9. Church rule

10. A good friend indeed

11. Dessert favorite

12. Arab overgarment

13. Funny one

21. Like some numerals

22. Insult, slangily

25. Set of steps

26. Redhead's secret

27. Islamic bigwig

29. Farthest or highest (Abbr.)

30. Marsupial pocket, e.g.

31. "Yippee!"

32. ___ of Nantes, 1598

33. DVD player button

37. NATO member

38. 'L'___ del Cairo" (Mozart opera)

39. 3 stanzas and an envoy

40. Wise legislator

42. Fire up

45. Marbles you don't play with

46. Blood pressure raiser

47. The organ of sight

48. The boss' "echo"

52. Stratified rock

53. Joins the choir

54. Bothersome burden

55. Hit prefix

56. Bog fuel

57. Egyptian cobra

58. Bookkeeper

59. Charged particle

PUZZLE 13

1	2	3	4		5	6	7	8	9		10	11	12	13
14					15						16			
17					18						19			
20				21						22				
23							24				25	26	27
			28		29	30	31			32			
33	34	35		36				37	38				
39			40		41					42			
43			44					45		46			
47						48				49			
50				51	52				53		54	55	56
		57				58	59	60					
61	62	63		64					65				
66				67					68				
69				70					71				

ACROSS

1. Dead-end jobs, e.g.

5. Bars or bolts

10. Shark variety

14. Adjoin

15. Danger

16. Decorated a cake

17. Speck of dust

18. Marine biology subject

19. Eye annoyance

20. Defender of some group or nation

23. Police trap

24. Versus

28. Drinks alcohol to excess habitually

32. Grassy plain of Latin America

33. Baseball great Mel

36. To get in or get out

39. Mercury and Saturn, but not Earth

41. Like poltergeists

42. There's probably money in it

43. 365 (or 366) days

46. Bloom-to-be

47. Very, to the maestro

48. Artifices

50. Ones with iron hands

53. Hitches successfully

57. Nonsense

61. Aerated beverage

64. Norse love goddess

65. German leader Helmut

66. 'A Prayer for ___ Meany"

67. Let go

68. Fanzine focus

69. Push a product

70. Feet are in them

71. McGee's closet, e.g.

DOWN

1. Boat launches

2. WWII sub

3. 'Frutti" intro

4. Some skeletal parts

5. Reach across

6. Tailor-made items

7. Fertilizer ingredient

8. Knee-to-ankle bone

9. It's the word on the street

10. Forcibly thrown or projected

11. Get off the fence

12. Anthem author

13. Shelley praise

21. Check out shamelessly

22. Powder substance

25. Mogul empire

26. Major mix-up

27. Trifled (with)

29. It might be skinned in the fall

30. Checklist unit

31. One way to get to first base

33. Eightsome

34. Leather strap

35. Portion of hair

37. Indigenous Japanese people

38. Clears, on a pay stub

40. Directly or immediately

44. Old orchestral string

45. For the missus

49. Former kingdom in northeastern India

51. It may be pulled at a carnival

52. Narrow channel

54. Semiconductor, perhaps

55. Distinctive spirit of a culture

56. Flies off the shelf

58. Title for Helmut Kohl

59. Word with green or eagle

60. X-ray units

61. One with a beat?

62. Symbol of wisdom

63. Money of Romania

PUZZLE 14

ACROSS

1. Person of exceptional holiness

6. Languages spoken

13. Fetched

16. Not of this world

17. Melanesia locale

18. Aussie avian

19. African grass

21. What you can spend

22. '... ___ nation under God ..."

23. Gloss target

24. Acronymic computer truism

25. Kathmandu native

29. Elapsed time

32. Hearty har-har

34. Janitor's implement

35. Backup cause, often

36. Animal on Michigan's state flag

37. Soybean container

38. 'That's all ___ wrote"

41. Flow stopper

43. It may be heard in a herd

45. Lunched or munched

46. Brewery unit

47. Connection of two or more computers or dedicated devices

52. Tram filler

53. Au --- (in gravy)

54. Recurrent twitch

55. Small amount

56. Espoused

57. Getty Museum pieces

58. Society girl

60. Notch

63. 'Now ___ seen it all!"

64. Election again

67. It's a hunting pooch

71. May neglect to

72. Greetings in old Rome

73. Stroke's need

74. "The One I Love" group

75. 'Ay, there's the ___ "

76. Natural disposition

83. Pub order

84. Thick-walled tubes

85. Stairway post

86. Xmas time

87. Vacation milieus

88. Minute

DOWN

1. Letters on a motor-oil can

2. The whole shooting match

3. It may be cast

4. Like a cool cat

5. Not at all sacred

6. '... on the ___ prairie"

7. Get off the fence

8. Missing leg

9. Priest's concern

10. Unfitting

11. Acapulco affirmatives

12. Contains

13. Magical wish-granter

14. Alpha's opposite

15. Aristotle, to Alexander the Great

20. Foolishly old-womanish

24. Swindle

25. Major broadcaster

26. Japanese delicacy

27. Heavily and firmly

28. Aquatic organisms

29. River to the Bering Sea

30. Wax rhapsodic, perhaps

31. Gnawing mammal

33. Not quite

37. Postal delivery

38. Crash helmet

39. Brood creator

40. One in a dozen

42. Yuman language

44. Frequently, poetically

47. Promise

48. Blood pressure raiser

49. Doctor, one would hope

50. Downy duck

51. Social stratum

59. 'Seinfeld" character

61. Charged particle

62. Smithereens maker

63. Tag antagonists

64. Venetian commercial center

65. More hair-raising

66. Dental anchor

67. Basic unit of capacitance

68. Seed source

69. Three-masted vessel

70. Customary functions

74. Sea in Antarctica

76. Prickly seed vessel (Var.)

77. Poet's "before"

78. PC screen, possibly

79. Genealogy word

80. It gets fleeced

81. Fractional monetary unit of Japan

82. Underhanded

PUZZLE 15

ACROSS

1. Country on the Arabian Sea

5. Materials for securing

11. One place to be lost

16. Word with china or spur

17. Sitting room?

18. Garb for grads

19. Vague and poorly defined

22. ___ and Gomorrah

23. Last degree

24. Heredity determinant

25. Throat bacterium, for short

26. British saloon

29. Backrub response

31. Ballet step

32. Over which that nation exercises sovereign jurisdiction

36. They're history

37. Succors

38. 'Okie From Muskogee" Haggard

39. 'Peter ___" (Disney film)

40. Butter unit

41. Lunched or munched

43. Billiards stroke

44. Easily set off, as a temper

47. Noted water conserver

50. 'You ain't seen nothin' ___!"

51. ___ about (wander)

52. Engine speed, for short

55. Some mites

56. Star in Orion

58. Desiccated

59. Nuclei of atoms; mediated by gluons

63. Carol contraction

64. Leg, in slang

65. They're worth three points in Scrabble

66. Harangues

67. Alimentary canal

69. Mineral spring

71. Thicket of trees

72. Start from the beginning

78. Bye, in France

79. Clears for takeoff, in a way

80. It may be heard in the Highlands

81. Pine

82. Stands for things?

83. Meat and vegetable dish

DOWN

1. ____-Wan Kenobi

2. Often grown as houseplants

3. Native or inhabitant

4. Wants or needs

5. Bud holder

6. Rocky ridge

7. Part of a basketball hoop

8. He worked on canvases

9. Historic Quaker

10. Existing with reference to, a State of the American Union

11. It's often lied about

12. Half a picker-upper

13. Employee who sweeps

14. Entangle or catch in

15. A person

20. Indian helmet

21. Freeze follower

25. 'That's one small ____ for ..."

26. Maya Angelou's forte

27. You may bookmark it

28. Involving two parts or elements

30. Exaggerate one's acting

33. Spicy type of food

34. Italian wine city

35. Dotted-line command

40. A fence made of upright pickets

42. Moth having nonfunctional mouthparts as adults

44. White knight, stereotypically

45. Seagull relative

46. Splashy party

47. Figure in many a New Yorker cartoon

48. Numbers 89 through 103

49. Newborn offspring are suckled

52. Implant again

53. Enzyme that catalyzes

54. Type of wear

57. It's placed in a setting

58. Basic commodities

60. Swordplay injury

61. Hinder

62. Gator cousin

68. Farsighted one

70. Not docked

71. "Pool" intro

73. 252-gallon unit

74. Fleur-de-____

75. 'Dead man's hand" card

76. ____ Aviv

77. Hot off the press

PUZZLE 16

ACROSS

1. "La Scala di ___" (Rossini opera)

5. Bunch

9. Get temporarily

16. Liszt's "La Campanella," e.g.

18. Shrek, e.g.

19. Island volcano

20. image has colored fringes

23. Ground cover

24. Automatic

25. "I ___ you!"

26. Conciliatory

28. ___ bread

30. ___ Today

32. One of Alcott's "Little Men"

33. Causing fear

40. Aces, sometimes

42. Trick taker, often

43. Certain surgeon's "patient"

44. Bygone bird

45. Chartered companies of London

50. Attracts

52. Hon

53. Milky gems

55. Auspices

56. It can come after a shot

59. Matterhorn, e.g.

60. Tart

62. ___ erectus

63. Drain

66. Siberian forest

68. Beside

70. African grass

75. Bird ___

76. Ship fitted out at Palos

78. Appear

79. "Trick" joint

80. Ahead of the times

86. Sylvester, to Tweety

87. "I ___ you one"

88. Bacchanal

89. Common European thrushes (Var.)

91. Bit

94. Indian helmet

96. Birdlike

97. installed under a kitchen sink to grind

104. Preserve a dead body

105. A head

106. Online currency

107. Some stanzas

108. Heavy cart

109. Abounding with elms

DOWN

1. "Hold on a ___!"

2. Addis Ababa's land: Abbr.

3. Russian writer

4. Beautify

5. Drunkards

6. Modern F/X field

7. Parenthesis, essentially

8. Scarcity

9. In need of resupply, maybe

10. "___ Gang"

11. Biochemistry abbr.

12. Wicker material

13. Assortment

14. Cashmere, e.g.

15. Beach, basically

17. In-box contents

19. Big Apple attraction, with "the"

21. Fungal spore sacs

22. Big pig

26. "American ___"

27. Doctor Who villainess, with "the"

28. River to the Rio Grande

29. Any thing

30. Adaptable truck, for short

31. Lentil, e.g.

34. Almanac tidbit

35. "Empedocles on ___" (Matthew Arnold poem)

36. Aromatic hydrocarbon derivative

37. Adult insect

38. Kind of situation

39. Bloated

41. Attacks

46. ___ v. Wade

47. Auld lang syne

48. Sea anemone, for example

49. Software program, briefly

51. Say

54. Buttonhole, e.g.

56. Refuse

57. Very loud utterance

58. French romance

59. Absorbed, as a cost

61. ___ roll

64. Darn, as socks

65. ___ du jour

66. Cheap

67. The "A" of ABM

69. Vex, with "at"

71. Put up, as a picture

72. Of or relating to antennae

73. Close, as an envelope

74. Adjusts, as a clock

77. Anger

81. Burrowing herbivorous Australian

82. Arizona Indian

83. Having colors

84. ___ Scotia

85. Cunning

90. Booze

91. Long, long time

92. Break

93. Eyes, poetically

94. Caddie's bagful

95. "That's ___ ..."

96. Far from ruddy

98. A pint, maybe

99. Clock standard: Abbr.

100. When it's broken, that's good

101. Mozart's "L'___ del Cairo"

102. An end to sex?

103. "___ will be done"

PUZZLE 17

1	2	3	4	5		6	7	8	9	10		11	12	13
14						15						16		
17				18						19				
20				21				22						
23			24		25		26			27				
28			29					30			31			
	32					33				34				
		35		36	37		38		39					
40	41	42		43		44			45		46	47		
48			49				50	51						52
53		54			55					56				
57				58			59		60		61			
62					63				64					
65				66					67					
68				69					70					

ACROSS

1. Dimension PCs, e.g.

6. Bridge support

11. Half of a lively dance

14. Form of electronic communication

15. Worn and torn

16. The Chiffons' "___ So Fine"

17. Lock

20. Lunched or munched

21. Recurrent twitch

22. Type genus

23. Rumor generator?

25. Basic monetary unit of Ghana

27. Wore

28. Cause (a compound) to polymerize

31. Semicircle, e.g.

32. Endeavour acronym

33. It's placed in a setting

34. Where starter

35. Thus

38. Song for Simon and Garfunkel, e.g.

40. Perfect scores

43. Black gold

45. Traffic sign word

48. Cell substance

49. Role

53. Message boat

55. Medical discovery

56. Mineral deposit

57. Devisal

59. Matador's trophy

61. Little tyke

62. Electric currents

65. Form compost

66. Even the score again

67. Japanese poetry

68. Finder of secrets

69. Reacted to fireworks

70. Brooding worry

DOWN

1. Pack up and leave

2. Hate, say

3. Thin membrane

4. Ad add-on

5. Notch

6. Inquiry for a lost package

7. Word in a supposed Cagney quote

8. Put to use

9. Promenade for Plato

10. Make synchronous

11. Relating to or consisting of

12. Browbeats

13. Sideways

18. Easy partner

19. Meadow, in verse

24. Disintegrates, as a cell

26. Archaeological excavation

29. ___ del Plata (Argentinean resort)

30. Aussie avian

34. Alcohol variety

36. Crib talk

37. Cloth

39. Announcement to passengers

40. Inventory of goods

41. Wrap

42. Sophistication or worldliness

44. Money of Romania

46. State of disorde

47. Move out of a dock

49. Speakers' spots

50. Took advantage of, with 'on'

51. Beef buyer's

specification

52. Reach or gain access to

54. 'Not my error" notation

58. Carpet-layer's concern

60. Pro follower

63. It may be cast

64. Supply with a staff

PUZZLE 18

1	2	3	4		5	6	7	8	9		10	11	12	13
14					15						16			
17				18							19			
20						21					22			
23				24		25			26					
		27			28				29			30	31	
32	33	34			35		36	37	38			39		
40				41							42			
43				44					45					
46			47				48		49					
		50			51	52	53		54			55	56	57
58	59				60			61		62				
63					64				65					
66					67						68			
69					70						71			

ACROSS

1. Way around London, once

5. Worrier's worry

10. Some chase scene maneuvers, slangily

14. Entire: Prefix

15. Who is't can ___ woman?': 'Cymbeline'

16. Tiny time unit: Abbr.

17. Trustee group at an Atlanta campus?

19. Fill in the blank with this word: ""___ never work!""

20. Fill in the blank with this word: "California's ___ National Forest"

21. Intestinal parts

22. Stars

23. Fill in the blank with this word: ""Vigilant ___ to steal cream": Falstaff"

25. In reverie

27. Some CBS forensic spinoffs

29. What online shoppers may spend

32. How can people ___ heartless?' ('Hair' lyric)

35. Old Scottish county on the Clyde

39. South Korea's Roh ___ Woo

40. Teacher's deg.

41. Decks out

42. Fill in the blank with this word: "England's Isle of ___"

43. War on Poverty agcy.

44. Zinger response

45. Tut's kin?

46. a habit worn by clerics

48. Tiger ___

50. Spasm

54. Once, but not twice

58. Remarked

60. Heyerdahl's second papyrus boat

62. The English translation for the french word: impression

63. Fill in the blank with this word: "___ cat"

64. Remembrance of things past

66. Fill in the blank with this word: "Also-___ (losers)"

67. Arrowsmith's wife

68. Unicorn in a 1998 movie

69. Video game island

70. Comparable to a rose?

71. Without restraint

DOWN

1. Trig angle symbol

2. Some tomatoes

3. Honolulu's ___ Tower

4. Home of the Atlas Mountains

5. City, informally

6. Start of a string of 13 Popes

7. Occult science

8. Conqueror of Northumbria in 946

9. What stealth planes avoid

10. The first commercially available electronic computer

11. Gauges

12. Squeal in pain

13. Looks unhappy

18. Sounds heard in passing?

24. the state of aggregation of soil and its condition for supporting plant growth

26. Squealed cries

28. They're taken in high sch.

30. Polio vaccine developer

31. Attention-getters

32. Fill in the blank with this word: ""___ good cheer!""

33. Schwalm-___ (German district)

34. Works with a plane

36. Writer Anais

37. Uzbekistan's ___ Sea

38. Fasten anew

41. Japanese golfer Isao ___

45. Percussion unit

47. Campus in Brookville, L.I.

49. Van Gogh's was "for life" according to Irving Stone

51. Start of a refrain

52. Starbucks stores

53. Fill in the blank with this word: "___-miss"

55. Rose-red dye

56. University mil. programs

57. Two-tone treats

58. W-2, e.g.

59. Fill in the blank with this word: "___ and done with"

61. Fill in the blank with this word: ""Able was ___...""

65. X-ray spec?

PUZZLE 19

1	2	3	4		5	6	7	8	9		10	11	12	13
14					15						16			
17			18						19					
20				21					22					
		23					24	25						
26	27	28				29					30	31	32	
33					34					35				
36				37					38	39				
40				41					42					
43			44	45				46						
		47				48								
49	50	51				52					53	54	55	
56				57	58				59					
60				61					62					
63				64					65					

ACROSS

1. Penpoints

5. unpleasant (similar term)

10. Fill in the blank with this word: "___-ran"

14. Words with a nod

15. Lash ___ of old westerns

16. To decide irrevocably, like someone's fate

17. Source of pop-ups?

20. Wanted-poster letters

21. Work unit

22. Je t'aime : French :: ___ : Spanish

23. Fill in the blank with this word: "___ Raton, Fla."

24. Formal hat, informally

26. Off-road motorcycle competition

29. Where flies get swatted?

33. Pursue, in a way

34. Sir Peter ___, painter of British royalty

35. Boatload

36. Sons of Leda

40. Clearasil target

41. Fill in the blank with this word: "___ dixit"

42. Fill in the blank with this word: "___ rubber (shoe material)"

43. Do some courtroom work

46. Squirrel, to 35-Down

47. Part of qq.v. (which see)

48. Senator Jake in space

49. Six-time baseball All-Star Rusty

52. Marshal ___, Yugoslavian hero

53. Fill in the blank with this word: "___-to"

56. Alcohol, it's said

60. Keto-___ tautomerism (organic chemistry topic)

61. Kentucky's ___ College

62. "That was close!"

63. The English translation for the french word: toge

64. Fill in the blank with this word: "___ 500"

65. Swiss watch brand

DOWN

1. Singer Simone

2. Writer Dinesen

3. Phi ___ Kappa

4. Variety of whale

5. The English translation for the french word: alnico

6. Occult science

7. Writer ___ Stanley Gardner

8. Was revolting

9. Fill in the blank with this word: "Comic actress ___ Lillie"

10. Star: Prefix

11. Sci-fi princess

12. With 69-Across, 1930s-'50s bandleader

13. You might stick a knife in it

18. Wine city north of Lisbon

19. Watch part

23. Fill in the blank with this word: "___ Ward, who played Robin in TV's "Batman""

24. What appears above a pi

25. Week or month at the office, usually

26. Auto racer Al

27. Tiring problem for bicyclists?

28. Name for a cowpoke

29. Thickheaded

30. N.Y. Jet or Phila. Eagle

31. Plum, for one

32. Poultry plant worker

34. The English translation for the french word: retomber

37. Pirate-fighting org

38. Old import tax

39. Scientology's ___ Hubbard

44. The Eagle in the heavens

45. Santiago de ___

46. Small-time dictator

48. Razzed

49. Fixed at an acute angle

50. Up ___ good

51. Slip ___ (err)

52. Wrench

53. Funny ___

54. Linear

55. *I.R.S. form

57. Weight abbr.

58. Untilled tract

59. You might save a life if you know this procedure combining chest compressions with mouth-to-mouth breathing

PUZZLE 20

ACROSS

1. Remit

5. Skinflint

10. Some mollusks

16. Healthy berry

17. Diminish in intensity

18. English accent

19. Not inclined

21. Pointed beard

22. Green fabric

23. Flush out

24. Passions

25. Public promotion

29. Clumsy dancer's obstacles

30. Nonkosher food

31. Organic compound

32. Ovoid objects, to Romans

35. '48 ___ " (Nolte-Murphy film)

36. Networked computers, for short

37. Dark (brown) hair

39. Bled

40. Type of shake

41. Scurries

42. Responses without delay

47. Willa Cather's "One of ___ "

48. Geishas' accessories

49. Charged particle

50. Quantity produced

53. Motion pic format

54. Blouse, e.g.

57. Mineral spring

58. Modest skirt length

59. It may be heard in a herd

60. Bowed, in music

61. Surgical removal

64. Ammonia molecules

67. Animal on Michigan's state flag

68. Some salad cheeses

69. Attach to stones or seaweed

70. Flat bone

72. Per follower

73. Alpha's opposite

74. Spanish liqueur

75. Foot part

76. Balance sheet plus

77. Min. components

DOWN

1. Holy day

2. Nation named for its location

3. Tour guides, often

4. Disorganized person

5. Supply with a staff

6. Peninsula in Europe

7. Enjoyment in cruelty

8. Chopin composition, e.g.

9. Standing on the street?

10. System of principles

11. Chuckle softly

12. Some amphibians

13. Kevin's "A Fish Called Wanda" role

14. Mooring spot

15. Tries to reduce swelling, in a way

20. Contraction in ''The Star-Spangled Banner''

26. Do an Oscar winner's job

27. Is worthy of

28. Accustom to hardship (Var.)

32. Singer Redding

33. Vivid twosome

34. The law, to Mr. Bumble

36. Anita Brookner's ''Hotel du ___''

37. Handiwork to dye for

38. At that time

39. Computer architecture acronym

40. Protein

42. Witty comment

43. Fertilizer ingredient

44. 'Cheers" character

45. Green damage

46. Assent gesture

47. Photo ___ (media events)

51. Come from

52. Intercept a conversation, in a way

53. Vessel ropes

54. Musical interval

55. Originating in the sea

56. Call one's own

59. Bench-clearing brawls

60. One-celled organisms

61. Maximum bet, e.g.

62. Cold carriers?

63. Frequently, poetically

64. Fungal spore sacs

65. Plebeian

66. Atlas contents

70. Its embrace is lethal

71. Assumed the lotus position

PUZZLE 21

ACROSS

1. Atlantic fish

5. Stays behind

12. Bean counter, for short

15. In shape

17. A sound between two words

18. Cabernet, e.g.

19. Come to mind

20. A poor landlocked country in western Africa

22. Incapable of being dissolved

24. Half a sawbuck

25. "The Matrix" hero

26. "It's no ___!"

27. Kiss

30. ___ dark space (region in a vacuum tube)

32. Staircase post

34. Flight data, briefly

35. A long-haired breed of cat

38. Losing proposition?

41. 30-day mo.

42. "The English Patient" setting

47. A republic in the West Indies

52. Florence dialect

53. Ring bearer, maybe

54. "O" in old radio lingo

55. A warm ocean current

60. Berry touted as a superfood

64. Beat

65. Dark area

68. Cover by strewing

70. Grand ___ ("Evangeline" setting)

71. "Gimme ___!" (start of an Iowa State cheer)

73. Certain Ècole

74. Tall flowering plant

77. Take, eat; this is my body,

81. English assignment

82. "My boy"

83. Break

84. Balkan capital

85. Atlanta-based station

86. Kind of penguin

87. "___ only"

DOWN

1. One may be taken to the cleaners

2. Eye bank donation

3. Accord

4. 100 centavos

5. Jewish teacher

6. Month preceding Rosh Hashanah

7. Bog

8. "Don't ___!"

9. "Rocky ___"

10. "Smoking or ___?"

11. Big mess

12. Tie

13. 100 centimos

14. Eats up

16. Deceive

21. Carpentry tool

23. ___-friendly

27. Control

28. "Hold on a ___!"

29. Catch

31. Kind of nerve

32. Basketry palm

33. Heartfelt

36. Amniotic ___

37. ___ cross

38. "Silent Spring" subject

39. Chit

40. Dash lengths

43. Big boomer

44. Priestly garb

45. "Flying Down to ___"

46. Trick taker, often

48. Bother

49. Accustom

50. Miles per hour, e.g.

51. "To ___ is human ..."

56. Blue

57. Good times

58. 100 cents

59. A cylindrical spikelike inflorescence

60. Least incompetent

61. British crown colony from 1802 to 1948

62. Rise

63. Any thing

66. Lessen the density or solidity of

67. Social breakdown

69. "The Canterbury Tales" pilgrim

70. Inquires narrowly and searches

72. "Don't get any funny ___!"

74. Apple spray

75. A chorus line

76. Loafer, e.g.

78. Game with matchsticks

79. Blouse, e.g.

80. A pint, maybe

PUZZLE 22

ACROSS

1. Asian nurse

5. Telekinesis, e.g.

8. Computer architecture acronym

12. Opera star

13. Five spots on a five-spot

14. Andrea Doria's domain

15. Soon, to a bard

16. Related to the anus

17. Synthetic resin

18. Spent only for a particular purpose

20. "Naked Maja" painter

21. Put to rest, as fears

22. Calendar abbr.

23. Stellar

26. Repudiate

30. Dash abbr.

31. Wicker material

34. Container weight

35. ...

37. ____ grass

38. Arctic

39. Brass component

40. Gooey cake

42. Chill (out)

43. Brainiac

45. Cause of hereditary variation

47. "Spy vs. Spy" magazine

48. Symmetric crystal

50. Cicatrix

52. Hoofed mammals having very thick skin

56. Recipe direction

57. A chip, maybe

58. Greek earth goddess: Var.

59. Bad for dieters

60. Animal with a mane

61. Catch a glimpse of

62. At the home of

63. Ring bearer, maybe

64. Bakery selections

DOWN

1. Jewish month

2. Peewee

3. Shakespeare, the Bard of ___

4. Phantom home?

5. Feather, zoologically

6. Out there

7. "Cast Away" setting

8. Set up new headquarters

9. Black

10. ___ bean

11. Blackguard

13. Spanish dish

14. Light green plums

19. Danger signal

22. Kipling's "Gunga ___"

23. Astound

24. Bit of parsley

25. Flip-flop

26. Anniversary, e.g.

27. Brass button?

28. Architectural projection

29. Piece cut from a cheese wheel

32. Contemptible one

33. Sylvester, to Tweety

36. Sentimentality in art or music

38. unning water

40. "Crikey!"

41. German cathedral city

44. Before the due date

46. Record holder?

48. Craze

49. Gibson, e.g.

50. Literally, "king"

51. Commend

52. Blanched

53. "Duck soup!"

54. Opportune

55. 1951 N.L. Rookie of the Year

56. A fluorocarbon with chlorine

PUZZLE 23

ACROSS

1. Flower girl's scatterings

7. Cookbook abbr.

11. Caught some Z's

16. Global feature

17. Sundae topper, perhaps

18. Punt propeller

19. They were once going places

21. About to explode

22. Bolted

23. "Fantasy Island" prop

24. Roman historian

26. Uncritically satisfied with oneself

30. "The Three Faces of ___"

33. Romanian money

34. Breakfast area

35. Husk

36. Good doctor's attribute

40. Ed.'s request

41. Complain

42. Head, for short

43. "Idylls of the King" character

45. Penetrated

47. "Are we there ___?"

49. Paperless reading

53. Egyptian fertility goddess

55. Cheat, slangily

57. Contradict

58. Chucklehead

61. Compound that contains four chlorine atoms per molecule

64. Band with the hit "Barbie Girl"

66. Mountain goat's perch

67. Dusk, to Donne

68. "That's ___ ..."

69. Resolute adherence to your own ideas or desires

73. Carter, Clinton or Gore

74. Brouhaha

75. Balcony section

79. Suffix with sect

80. An increase in seriousness

84. Link

85. Icky stuff

86. Bolster or strengthen

87. Classmate of Ron and Hermione

88. Christmas season

89. Fancy

DOWN

1. Ancient Briton

2. Bounce back, in a way

3. Binge

4. High points

5. Family dog, for short

6. Relieved

7. Issue

8. "My man!"

9. Undertake, with "out"

10. Put off

11. Brightest star in Virgo

12. A hard protective sheath

13. Any of various widely distributed beetles

14. Porky's love

15. Railroad support

20. Do a pit stop chore

25. Unsaturated hydrocarbon

27. Snob

28. Typically

29. Back-to-work time: Abbr.

30. Declines

31. The Sail (southern constellation)

32. "Our Time in ___" (10,000 Maniacs album)

37. "The Open Window" writer

38. ___ Day

39. Any long object resembling a thin line

44. Live wire, so to speak

46. Clan members

48. ___-tac-toe

50. Assortment

51. Choreographer Michael

52. Lentil, e.g.

54. Approach

56. Carbolic acid

58. Holy day

59. Glass-walled enclosures

60. Like a pulp or overripe

62. "Awesome!"

63. Downwind

65. Star in Aquila

70. Hybrid offspring of a male

71. Golden Hind captain

72. Gunk

76. "My bad!"

77. Lions' prey

78. Catch a glimpse of

81. Former French coin

82. Newspaper div.

83. Toni Morrison's "___ Baby"

PUZZLE 24

1	2	3	4	5		6	7	8	9		10	11	12	13
14						15					16			
17						18					19			
			20		21					22				
23	24	25							26					
27					28	29	30	31						
32					33									
34					35						36	37	38	39
			40						41					
	42	43						44						
45	46							47						
48				49	50	51	52							
53				54				55		56	57	58		
59				60				61						
62				63				64						

ACROSS

1. ___ salad

6. Rank above marquis

10. "Aeneid" figure

14. Andrea Doria's domain

15. Component used in making plastics and fertilizer

16. "I, Claudius" role

17. Dead to the world

18. Long, long time

19. Indian dish made with stewed legumes

20. Animals having saclike bodies

23. Web-footed mammals

26. Magical characters

27. Share of the net profits of the business

32. Buenos ___

33. Graceful bird

34. "___ of the Flies"

35. Carried on

36. Hamster's home

40. Arcade coin

41. AprËs-ski drink

42. A teacher in a school below the college level

45. Norse goddess of love

47. Leftovers

48. Of or relating to lines of longitude

53. "Once ___ a time..."

54. Not "fer"

55. Increase

59. ___ mortals

60. Coagulate

61. Indian yogurt dip

62. At one time, at one time

63. Apprentice

64. Macho guys

DOWN

1. Modern F/X field

2. Biddy

3. Charlotte-to-Raleigh dir.

4. Feed someone who will not or cannot eat

5. A state of sudden spiritual enlightenment

6. Kind of income

7. Advocate

8. Astute

9. "___ of Eden"

10. Football play

11. Move, as a picture

12. Be bombastic

13. Undersides

21. "C'___ la vie!"

22. "___ Brockovich"

23. "How ___ Mehta Got Kissed, Got Wild, and Got a Life" (Kaavya Viswanathan novel in the news)

24. Beethoven's "Archduke ___"

25. Unit of pressure

28. Military cap

29. Philosophy of Right' writer

30. "You ___ kidding!"

31. Pistol, slangily

35. Court

36. The snail-shaped tube

37. Advil target

38. Attendee

39. All ___

40. Asian tongue

41. Inflammation of the nose

42. Men of La Mancha

43. Young swan

44. Long, long time

45. Logging channel

46. Cowboy

49. Boor's lack

50. Hideous

51. Christian name

52. "What's gotten ___ you?"

56. Animation

57. Adaptable truck, for short

58. Calphalon product

PUZZLE 25

1	2	3	4		5	6	7	8	9		10	11	12	13	14	15
16					17						18					
19			20								21					
22					23						24					
25				26			27	28								
29					30			31					32	33	34	
35				36			37					38				
		39			40						41					
	42	43			44				45	46					
47					48				49							
50			51	52				53				54	55	56	
57			58				59			60					
		61				62			63						
64	65	66				67			68						
69					70			71							
72					73					74					
75					76					77					

ACROSS

1. "Star Trek" speed

5. Card game for three

10. Muslim decrees

16. Fishing, perhaps

17. Guts

18. Shirker's excuse

19. Hearing by seeing?

21. An American in Paris, maybe

22. Work, as dough

23. ___ be an honor'

24. Free money?

25. Shop with the eyes only

29. A chip, maybe

30. ___ v. Wade

31. Doctor Who villainess, with "the"

32. Cabernet, e.g.

35. "Absolutely!"

36. "Gee whiz!"

37. Fragrant black nutlike seeds

39. Same old, same old

40. "___ on Down the Road"

41. Pro ___

42. English composer

47. Curve

48. Gigantic

49. "... ___ he drove out of sight"

50. A woman who is engaged to be married

53. "___ Maria"

54. Balaam's mount

57. "60 Minutes" network

58. Cover

59. "I ___ you one"

60. Ancient Andean

61. Never despair

64. Like tears

67. ___ el Amarna, Egypt

68. Spot broadcast, often

69. World Heritage List maintainer

70. A storm in which violent winds are inb hot air

72. Beneficiaries

73. Intimate

74. Length x width, for a rectangle

75. Tormentor

76. Perfume

77. Hires competition

DOWN

1. A path set aside for walking

2. Fatuous

3. Wishes to undo

4. Memorial Day event

5. Egg cells

6. Avenue east of Fifth

7. Cheerful

8. Sonata section

9. .0000001 joule

10. Devils' playground?

11. Transylvanian Alps setting

12. Imitating

13. Kind of life

14. 100 cents

15. Check

20. Tokyo, formerly

26. Anger

27. "To your health!"

28. Control ___

32. Enlarge, as a hole

33. Beanery sign

34. ___ lab

36. Kind of shot

37. Leather strap

38. Highlands hillside

39. Paul of 'Dinner for Schmucks'

40. Allow

42. Hop, skip or jump

43. Cuckoos

44. Bypass

45. Part of a voting machine

46. Anger

47. "Monty Python" airer

51. Displays clearly

52. Bony fish

53. Irreverent

54. Its official language is Catalan

55. Remove the scum from

56. Delhi dishes

59. A segment of DNA containing adjacent genes

60. Bit of progress

61. Japanese-American

62. Attendance counter

63. Appropriate

64. Brewski

65. Dwarf buffalo of Indonesia

66. Advance

70. TV monitor?

71. "Comprende?"

PUZZLE 26

ACROSS

1. Cig

6. Bad day for Caesar

10. The Sail (southern constellation)

14. Leg bone

15. Astronaut's insignia

16. Persia, now

17. About

18. Assign great social importance to

20. Mayor of a municipality in Germany

22. Baby's first word, maybe

23. "Seinfeld" uncle

24. "What's ___?"

25. "... ___ he drove out of sight"

26. Addition

27. "Empedocles on ___" (Matthew Arnold poem)

29. Boris Godunov, for one

31. "Mi chiamano Mimi," e.g.

32. In favor of

34. An organization of missionaries in a foreign land

37. Operating on living animals

39. Crusader's foe

40. www.yahoo.com, e.g.

41. ___ function

42. Jewish month

44. Charged particles

48. Bank offering, for short

49. Amigo

51. "Silent Spring" subject

53. Brouhaha

54. Antiquity, in antiquity

55. Containing or resembling amethyst

58. Action takeing place during a road journey

60. Up, in a way

61. Baptism, for one

62. Bad look

63. Cooktop

64. Again

65. "... or ___!"

66. Positions

DOWN

1. It'll hold your horses

2. Ballroom dance

3. Moon of Uranus

4. Double-decker checker

5. Big name in stationery

6. Marriage acquisition

7. Honoree's spot

8. Abstruse

9. More rational

10. Relative of "i.e."

11. Cosmopolitan genus of usually perennial herbs

12. Ointment ingredient

13. Neighbor of Namibia

19. Anger

21. Kind of unit

28. All excited

30. Change, as the Constitution

31. Garlicky mayonnaise

33. Egg cells

35. Well-built

36. "Dear" one

37. A salt or ester of vanadic acid

38. It rises every year

39. THEME ANSWER 5

41. Code word for "S"

43. Cleave

45. Familiarize

46. Minority

47. Stockholm natives

49. "Polythene ___" (Beatles song)

50. Plant used as soap

52. Autocrats

56. Associations

57. Asian tongue

59. Drops on blades

PUZZLE 27

ACROSS

1. Agenda entries

6. Dresses

12. Be itinerant

16. Water wheel

17. "Four Essays on Liberty" author Berlin

18. "What've you been ___?"

19. A disposition to be sly and stealthy

22. Aviation

23. "Thank You (Falettinme Be Mice ___ Agin)" (#1 hit of 1970)

24. 1,000-kilogram weights

25. Sylvester, to Tweety

26. Cow, maybe

27. "My man!"

30. Pass on

31. Not able to be ascertained

35. "Rocks"

36. On in years

37. Antares, for one

40. Adult

43. Parenthesis, essentially

44. Pickpocket, in slang

46. "For shame!"

47. A woman's undergarment

51. ___ Tuesday (Mardi Gras)

52. "C'___ la vie!"

53. Born, in bios

54. Propelled a boat

55. E.P.A. concern

57. Native American tents (Var.)

59. Arch

60. Covered wagon

66. "A Doll's House" playwright

68. "Star Trek" rank: Abbr.

69. Fix, in a way

70. Bauxite, e.g.

71. People of the Yucat√°n

73. Always, in verse

74. "___ the night before ..."

75. Market forces

80. Coastal raptor

81. Fermented Middle East beverage

82. Change, as a clock

83. Endurance

84. Presentation aids

85. Certain hearings

DOWN

1. Undisturbed

2. Colorful bird

3. Slips

4. Catalan painter Joan

5. Ed.'s request

6. Nazi

7. "___ I care!"

8. Dracula, at times

9. "Rocky ___"

10. Chinese "way"

11. ...

12. A small stream

13. Post for all to read

14. Baffled

15. Antiquated

20. Flycatcher

21. 'Ad" or "ab" ending

26. Bow

27. Wall Platform

28. Free from, with "of"

29. Antsy

32. Communicate silently

33. Spelling of "Beverly Hills 90210"

34. "Mi chiamano Mimi," e.g.

38. Bang-up

39. Abbr. after many a general's name

40. Ices

41. Froth

42. Post-mortem

43. Clothing

45. Opening words

48. ___ Spumante

49. 1987 Costner role

50. ___ Clinic

56. Animal fat residue

58. A.T.M. need

59. Bake, as eggs

61. "___ and the King of Siam"

62. Pond feeders

63. Not at all

64. Desk item

65. Turns back, perhaps

66. Drive

67. Wilkes-___, Pa.

72. Alone

73. Carve in stone

74. Deuce topper

76. "To ___ is human ..."

77. Victorian, for one

78. 40 winks

79. Ace

PUZZLE 28

ACROSS

1. Italian ice cream

8. Strong liquors

15. Hunks

16. Approach with stealth

17. Separated

18. Apply again

19. Lover with a ladder, perhaps

20. John and others

21. M√°laga man

22. Hires competition

23. Blue hue

25. Twangy, as a voice

27. Threadbare

28. Beyond calculation or measure

33. "Rocky ___"

34. Barely beat

35. Resolve

36. "Smoking or ___?"

37. In-flight info, for short

38. Expose to cool or cold air

40. Victorian, maybe

42. Accustom

43. Western Samoa money

44. [Just like that!]

46. Freshwater fishes

50. Giving the sensation of tension

52. Outcast

53. Corrupt morally or by intemperance or sensuality

54. Rap variety

55. Figure

56. Bromo ingredient

57. Eternal

58. Class work

DOWN

1. Checks out

2. Big name in computers

3. Sage

4. Comfy shoes

5. To cure or restore

6. ___ de force

7. Death on the Nile cause, perhaps

8. New England catch

9. Salad green

10. Bucket of bolts

11. Catmint

12. A cocktail of vodka

13. A sharp transient wave in the normal electrical state

14. 007, for one

20. White meat mold

22. The Talented Mr. Ripley' star

24. Fabricator's forte

26. "Your majesty"

27. Any agent that retards

28. All thumbs

29. Treat with nitric acid

30. Shoulder board (Var.)

31. Spring sound

32. Corker

39. Mob disperser

41. 1973 Elton John hit

44. Calamine targets

45. Certifies

47. Kid's name

48. Dine at home

49. Some food fishes

51. Dressing ingredient

52. French door part

53. ___ Appia

54. "For Me and My ___"

PUZZLE 29

ACROSS

1. Boutique

5. Big dipper

10. Boards

16. "I, Claudius" role

17. Arab leader

18. Cinch

19. In an Italian style

21. Spooks

22. Place for a barbecue

23. .0000001 joule

24. Minimum

25. Explodes when struck

29. Ashtabula's lake

30. Absorbed, as a cost

31. Advocate

32. Marienbad, for one

35. Burden

36. Chester White's home

37. Feeling nausea

39. Blue hue

40. ___ souci

41. Court attention-getter

42. Covered wagon

47. Drudgery

48. Neural network

49. Deception

50. A state of misfortune or affliction

53. "Cool" amount

54. Software program, briefly

57. "It's no ___!"

58. Household chore

59. Kills the helper T cells

60. Algonquian Indian

61. Circular firework

64. A complex inorganic compound

67. Charlotte-to-Raleigh dir.

68. Garlicky mayonnaise

69. Haitian monetary unit

70. Makes wet and dirty, as from rain

72. "The Wizard of Oz" prop

73. Of the vascular layer of the eye

74. Boat propellers

75. Extracts

76. Actress Oberon

77. European language

DOWN

1. Scrap

2. A female paramour

3. A woman plaintiff

4. Watch over

5. Grassland

6. Forgiveness of a sort

7. Honeybunch

8. "Take your hands off me!"

9. "... ___ he drove out of sight"

10. Blowhards

11. Coop flier

12. Autocrats

13. "No problem!"

14. Assayers' stuff

15. Home, informally

20. Chit

26. Debaucher

27. Shade

28. Infatuation

32. Eye affliction

33. Equal

34. Cutting tool

36. Schuss, e.g.

37. Foul

38. Bang-up

39. Ad headline

40. Boil

42. Miniature sci-fi vehicles

43. Bank of Paris

44. Dander

45. Antipasto morsel

46. Black gold

47. ___ cross

51. Kigali resident

52. Glossy fabrics

53. 20 Questions category

54. Like an iris part

55. Kitchen gadgets

56. A sleeveless cape with fur

59. Impede

60. Small tropical flea

61. About

62. "The Canterbury Tales" pilgrim

63. Card

64. All excited

65. Churn

66. Stubborn beast

70. Depress, with "out"

71. A pint, maybe

PUZZLE 30

1	2	3	4		5	6	7	8		9	10	11	12	13
14					15					16				
17					18					19				
20			21			22		23			24			
		25			26			27		28				
29	30	31						32				33	34	
35				36		37	38			39				
40				41							42			
43			44	45						46				
47			48				49		50					
		51				52								
53	54			55		56			57		58	59	60	
61			62			63		64			65			
66					67					68				
69					70					71				

ACROSS

1. "Hogwash!"

5. Neighbor of Libya

9. Approval

14. Halftime lead, e.g.

15. Exude

16. Throat dangler

17. Long, long time

18. Advertising sign

19. CÈzanne contemporary

20. Tech-heavy stock exchange

22. Dittography, e.g.

24. Decide to leave, with "out"

25. Bring out

27. Barely gets, with "out"

29. Draw into an argument

32. A ravine or gully in southern Asia

35. Functions

36. Transparent sea creature

39. Buddy

40. Same old, same old

41. Spin

42. ___ populi

43. And others, for short

45. Money in the bank, say

46. Greasy

47. Sewer line?

49. Examine

51. Bindle bearer

52. Jack

53. Chester White's home

55. Sundae topper, perhaps

57. Empathize

61. Any bird of the genus Pitta

63. Earned

65. Decorated, as a cake

66. Aromatic solvent

67. Hip bones

68. Abandon

69. Like some stadiums

70. Makes it

71. Atlantic City attraction

DOWN

1. Economical

2. "I had no ___!"

3. Freudian topics

4. Someone who transmits a message

5. *Balboa, e.g.

6. Clod chopper

7. Nitrogen

8. Contradict

9. Puzzle

10. "___ Maria"

11. Chetnik's country

12. High-five, e.g.

13. Brewer's equipment

21. "Much ___ About Nothing"

23. Rectangular area in front of the goal

26. Grabbers

28. "A Nightmare on ___ Street"

29. "Snowy" bird

30. Impudence

31. Brainwave in the encephalogram

33. Bikini, e.g.

34. A compound radical

37. Sue Grafton's "___ for Lawless"

38. Be exultant

44. "Seinfeld" uncle

46. Cleopatra's Needle, e.g.

48. On the train

50. "The Three Faces of ___"

52. Aussie "bear"

53. Didn't dillydally

54. Josip Broz, familiarly

56. Give off, as light

58. Advil target

59. Freshman, probably

60. Corm of the taro

62. Athletic supporter?

64. Archaeological site

PUZZLE 31

ACROSS

1. Bohemian, e.g.

5. Accomplishment

9. Permeate

14. ___ of the above

15. Annul

16. ...

17. From a foreign country

19. Lid or lip application

20. See if or how it works

21. National Zoo favorites

22. Nod, maybe

23. Reprimand, with "out"

24. Yellowish pink

28. Butt

29. "Awright!"

33. Rainbow ___

34. Exude

35. Ballad

36. Normal temperature of room

40. Person in a mask

41. Medical advice, often

42. Deceived

43. Poet Angelou

45. "Act your ___!"

46. Lead source

47. ___ Verde National Park

49. Keep out

50. Advantages

53. Soup cooked in a large pot

58. Balloon probe

59. A windstorm

60. Geometrical solid

61. On the safe side, at sea

62. History Muse

63. Anatomical dividers

64. Frau's partner

65. Suspended

DOWN

1. Arrogant and annoying

2. Look angry or sullen

3. The "A" of ABM

4. A constellation in the southern hemisphere

5. Charity event in the park

6. Provide, as with a quality

7. Gulf of ___, off the coast of Yemen

8. Anderson's "High ___"

9. Acquired relative

10. For the most part

11. Fasten

12. Component used in making plastics and fertilizer

13. All ___

18. "Jo's Boys" author

21. Cell alternative

23. "Carmen" composer

24. Play, in a way

25. Bouquet

26. Eccentric

27. Ornamental flower, for short

28. Small woods

30. Avoid

31. Composer Copland

32. Howler

34. Alpha's opposite

37. Clear, as a disk

38. Dismays

39. ___ el Amarna, Egypt

44. During

46. Elastic

48. Swelling

49. More despicable

50. Express Mail org.

51. Dermatologist's concern

52. Barber's motion

53. Stubborn beast

54. Allergic reaction

55. Yellowish brown balsam

56. "___ Brockovich"

57. E.P.A. concern

59. Code word

PUZZLE 32

ACROSS

1. Go ___ (fizzle)

5. Fill in the blank with this word: "___ bird"

9. Misbehaves

14. City on the Gulf of Aqaba

15. Truncation indications: Abbr.

16. Tiring problem for bicyclists?

17. Tell ___ story

18. Very bad, slangily

19. Stuffy sort, in slang

20. [1938]

23. Ticket abbr.

24. Wasn't colorfast

25. U.S.D.A. part: Abbr.

28. Tucson sch.

31. They make connections

36. The savory silver salmon with its bright red flesh is also known by this 4-letter name

38. Fill in the blank with this word: "___ vault"

40. Shinto shrine gate

41. Tennyson's 12-poem series

44. Where Hercules strangled a lion

45. Novice: Var.

46. ___ Institute (astronomers' org.)

47. Musician with the first record formally certified as a million-seller

49. 1936 Olympics hero

51. The Gateway to the West: Abbr.

52. ___ Precheck

54. S.A.S.E., e.g.

56. Advice for the impulsive consumer

65. Show, but not premiere

66. The English translation for the french word: zÈzayer

67. Italian composer Nino ___

68. Fill in the blank with this word: "___ wrench"

69. Wells's oppressed race

70. The English translation for the french word: osseux

71. Whiskey drinks

72. Adjust, as a clock

73. Look to ___ troublous world': 'Richard III'

DOWN

1. Fill in the blank with this word: "___ moss"

2. They take the bait

3. Ice ___

4. Tunable drum

5. Condensation

6. Buckwheat's affirmative

7. Year Justinian II regained the throne

8. Schindler of "Schindler's List"

9. total and all-embracing

10. Fill in the blank with this word: "___-chef (kitchen #2)"

11. Cause for an appointment with a cardiologist

12. Presidential ___

13. W.C.T.U. members

21. Indonesia's ___ Islands

22. Slangy turndown

25. More than passing

26. Mathematician Kurt

27. What each of the longest words in 17A, 65A, 10D and 25D famously lacks

29. When you pick up a tab, you do this to "the bill"

30. "Cavalleria Rusticana" baritone

32. Some Asian fighters

33. Ohio natives

34. Title role in a 1950s TV western

35. Magical symbol

37. Pulitzer-winning author Robert ___ Butler

39. Sein : German :: ___ : French

42. Signals for Revere

43. Amore from the Beatles, 1968

48. Misery

50. Nestle's ___-Caps

53. One who hears "You've got mail"

55. Reins in

56. They often precede la's

57. Realtor's specialty, for short

58. Manchurian border river

59. Stand for the deceased

60. Vex

61. Those, to Tom

62. Rude audience member

63. The ___ Reader (magazine)

64. Fill in the blank with this word: ""Divine Secrets of the ___ Sisterhood""

PUZZLE 33

1	2	3	4		5	6	7	8			9	10	11	12
13					14				15		16			
17				18						19				
20				21					22					
		23					24							
25	26	27			28	29					30	31	32	
33					34						35			
36				37						38				
39			40						41					
42		43				44								
		45				46								
47	48	49				50	51			52	53	54		
55				56				57						
58				59				60						
61				62				63						

ACROSS

1. Unskilled writer

5. The English translation for the french word: guÊde

9. Actress Lollobrigida

13. Torment

14. Unpaid debt

16. Wink ___ eye

17. Gull

20. With 22-Across, fourth in a series of five TV personalities (1992-2009)

21. Princess loved by Hercules

22. Nimrod

23. Grinder

24. Tony's portrayer on "NYPD Blue"

25. MAC

33. Totally rules

34. Stravinsky's "___ for Wind Instruments"

35. To the ___ power

36. Fill in the blank with this word: ""We'll give a long cheer for ___ men" ("Down the Field" lyric)"

37. Film extras, for short

38. Opera's ___ Te Kanawa

39. Fill in the blank with this word: "Dryden's "___ for Love""

40. Wordless song: Abbr.

41. St.-Tropez's Place des ___

42. Singers Starr and Kiki look at each other

45. "This Gun for Hire" star

46. Turn signal dirs.

47. W.W. II title

50. Tiny fraction of a min.

52. What a patrol car might get, for short

55. Tongue twister #3

58. Frosh, next year

59. Sides of some ancient temples

60. City in Judah

61. Plea

62. Fill in the blank with this word: ""Here I ___ Worship" (contemporary hymn)"

63. Finish this popular saying: "He who hesitates is_____."

DOWN

1. Trip to Mecca

2. The rain in Spain

3. Unscramble this word: cpoy

4. Show obeisance

5. University of Virginia players, familiarly

6. Turgenev's birthplace

7. Vissi d'___' (Puccini aria)

8. Wagner's "___ fliegende Hollander"

9. Wuthering Heights' genre

10. Fill in the blank with this word: ""___ not back in an hour...""

11. Touch alternative

12. Seraph of S

15. Wagga Wagga residents

18. South Pacific carvings

19. The English translation for the french word: insuffisant

23. Library section

24. There are lots of eaters of chocolate bunnies on this holiday

25. With 65-Across, go against the group ... or what the circled squares literally do in the answers to the starred clues

26. University of Missouri locale

27. Without warmth

28. Put up

29. Sales slips: Abbr.

30. Standing by

31. Up ___ (trapped)

32. At once

37. Relatives of TV host Tom

38. Youngsters

40. Call into question

41. Lascivious type

43. Like some winter sidewalks

44. Relative of a 29-Down

47. Fill in the blank with this word: "___ dixit"

48. Scientology's ___ Hubbard

49. State, e.g.: Abbr.

50. Fill in the blank with this word: "___-Penh"

51. Trick-taking game

52. The Ponte Vecchio crosses it

53. They have open houses

54. Rec rm. locale, often

56. Tony-winning Hagen

57. Drive forward

PUZZLE 34

ACROSS

1. Seating sect.

5. Fill in the blank with this word: ""Maria ___," Jimmy Dorsey #1 hit"

10. Fill in the blank with this word: "Den ___, Nederland"

14. Cosmos star

15. Tropical roots

16. Soprano Berger

17. Union foe

18. Modern viewing options, for short

19. Fill in the blank with this word: "Famous ___"

20. Protection in the city?

23. Year in Leo IX's papacy

24. Sigur ___ (Icelandic post-rock band)

25. Fill in the blank with this word: "Chekhov's "Uncle ___""

26. Oilman Kashoggi

28. Local theater, slangily

31. Troublemaker

32. Welsh symbol

33. Time and temperature, e.g.

36. Bridge

41. Literally, 'a blowing out'

42. Fill in the blank with this word: "Caladryl: itch :: Bengay : ___"

43. Middle of this century

46. Old-time actor Wallace ___

47. Cleaving tools

48. This single-celled fungus can ferment sugars & carbohydrates

50. Team VIPs

52. Fill in the blank with this word: "___ in Thomas"

53. Jacket for a hunting dog?

58. Fill in the blank with this word: "Barbara Kingsolver's "___ Am"

59. What detectives follow

60. Golfer's challenge

62. Trompe l'___

63. Fill in the blank with this word: ""Ich ___ dich" (German words of endearment)"

64. Fill in the blank with this word: ""___ Pastore" (Mozart opera)"

65. Waiting aid

66. Wrapped up

67. Unscramble this word: tsco

DOWN

1. Special ___

2. Happen again

3. The Music Man'S Instrument : Benny Goodman

4. "Sunny" singer Bobby

5. ___ jazz (fusion genre)

6. Hollywood's Alan and Diane

7. Ziegfeld Follies designer

8. Fill in the blank with this word: "De ___ (again)"

9. Made a tax valuation: Abbr.

10. Fill in the blank with this word: ""Do I ___ second?""

11. Valentino rival

12. Unidentified person

13. Valve in some fireplaces

21. Son of Prince Valiant

22. visible (similar term)

23. Fill in the blank with this word: "___ fide"

27. Similar (to)

28. Young dragonfly

29. Josephine Tey investigator ___ Grant

30. Fill in the blank with this word: "___ Cynwyd (Philadelphia suburb)"

33. Your majesty'

34. Year in Severus's reign

35. Zoom

37. Scottish inlet

38. Cactuslike tree of the Southwest

39. They're usually dark on Monday nights

40. Fill in the blank with this word: "Amerada ___ (petroleum giant)"

43. "Yeah, r-i-i-ight!"

44. Limo driver in the airport, e.g.

45. Highest worship in Catholicism

47. Camera setting

49. Reptilian, in a way

50. Fill in the blank with this word: ""___ with you" (parting words)"

51. Pondered

54. Fill in the blank with this word: ""Winnie ___ Pu""

55. Teutonic turndown

56. Vexed

57. Savings acct. protector

61. Fill in the blank with this word: ""Did you ___ that?""

PUZZLE 35

ACROSS

1. Willy Wonka's creator

5. Puerto ___, Chile

10. Where Jakarta is

14. Inter ___

15. Marie Osmond's ___ Belle dolls

16. Sixth Jewish month

17. Veg out

18. This is either a regional dialect or an odd phrase that can't be understood by looking at the individual parts

19. Bullfighter's cloak

20. Wit in need of washing?

23. Wanton look

24. Morning music

25. Tracy's "Tortilla Flat" co-star

28. Yom Kippur service leader

30. Marie Antoinette, e.g.

31. They're heard when Brits take off

32. Fill in the blank with this word: "Electric ___"

35. toward the mouth or oral region

36. Skewered Asian fare

37. Irving Bacheller's "___ Holden"

38. Sue Grafton's '___ for Noose'

39. The English translation for the french word: sahib

40. Fill in the blank with this word: ""___ say it is good to fall": Whitman,

"Song of Myself""""

41. Transmitter starter?

42. The English translation for the french word: attaquer

43. African fly

46. June 6 1944 is____.

47. Part of a hidden agenda: 2 wds.

52. Fill in the blank with this word: "___, zwei, drei"

53. What las novelas are written in

54. Roman statesman ___ the Elder

56. Heyerdahl's second papyrus boat

57. The English translation for the french word: pÈdoncule

58. Fill in the blank with this word: "___ out a living (scraped by)"

59. 1953 Oscar-nominated film based on a novel by Jack Schaefer

60. Series of online posts

61. What to click after finishing an email

DOWN

1. The Mavericks, on scoreboards

2. Southwestern trees

3. Sword handle

4. Space cadet's place

5. Longfellow or Millay, by birth

6. Stranger

7. Reason to be barred from a bar ... or the theme of this puzzle

8. Fill in the blank with this word: "Drop ___"

9. Team that has a tankful of rays in the back of its ballpark

10. Derek who played Claudius in "I, Claudius"

11. "Battlestar Galactica" commander

12. Uninteresting

13. Fill in the blank with this word: ""It's ___ against time""""

21. Fill in the blank with this word: ""Able was ___...""""

22. Wrigley team

25. Scientology's ___ Hubbard

26. Gas: Prefix

27. War stat

28. Unscramble this word: itrao

29. Run up ___ (owe)

31. Asian goat

32. Ancient city with remains near Aleppo

33. Webb Pierce song "___ Know Why"

34. Keto-___ tautomerism (organic chemistry topic)

36. Most impertinent

37. Classic 30's-40's radio comedy

39. Print tint

40. Fill in the blank with this word: ""Time ___ a premium""""

41. Unlikely Scottish sight

42. Slaphappy, say

43. Fill in the blank with this word: ""The Strange Love of Martha ___" (1946 film)"

44. West Coast wine city

45. Fill in the blank with this word: ""Another ___, Another Show" ("Kiss Me, Kate" song)"

46. Words with line, hint or bomb

48. Fill in the blank with this word: ""Scrubs" co-star ___ Braff"

49. Wood protuberance

50. Twister's trail

51. Coordinate in the game battleships

55. How ___!'

PUZZLE 36

ACROSS

1. Sports players: Abbr.

5. Like Christmas in Madrid?

9. Vowel sound

14. Zebulon Pike got only this glimpse of this, his namesake

15. Rate ___ (be perfect)

16. Campus buildings

17. Kuwaiti bigwig

18. Where Manhattan is: Abbr.

19. Wolf pack member

20. High water?

23. Some red giants

24. Sen. Gaylord ___, who originated Earth Day

28. the compass point that is one point south of due east

29. Old bird

31. Part of a Spanish play

32. What traffic and dogs do

35. St.-___, capital of R

37. Work started by London's Philological Soc.

38. Viking stories, e.g.

41. Groove-billed ___

42. Fill in the blank with this word: "___ Clark who sang "Poor, Poor Pitiful Me""

43. Polo and others

44. Swimming place

46. Fill in the blank with this word: "Cardio : heart :: ___ : ear"

47. Wayne LaPierre's org

48. Williams in the water

50. Whipped

53. A capital Nobel Prize novelist?

57. 1988 Peter Allen musical

60. Fill in the blank with this word: ""Nothing beats ___" (beer slogan)"

61. Sweat spot

62. Western Pacific republic

63. Fill in the blank with this word: "___ rock (radio format)"

64. Teardrop site

65. Precious ___

66. Specialty oven

67. 'Waiting for the Robert ___'

DOWN

1. Fill in the blank with this word: "Be ___ to (help out)"

2. U.S.S. ___, first battleship to become a state shrine

3. Unskilled writers

4. Picnic pests

5. Walkers on hot coals

6. The Louvre's Salles des ___

7. Filmmaker Riefenstahl

8. Rare book dealer's abbr.

9. Worcestershire ___

10. Secret doctrine

11. Wellness grp.

12. New Deal program inits.

13. Writer's helper: Abbr.

21. Writing pad

22. Have ___ of mystery

25. Wash abrasively

26. Saturday Night Live' alum Cheri

27. Knots

29. Subway

30. The ___ Love' (R.E.M. hit)

32. Potter professor Severus ___

33. Things to avoid

34. Running wild

35. What a vacuum cleaner vacuums

36. Small-time dictator

39. Arrowsmith's wife

40. Service station?

45. Like the Dalai Lama, historically

47. Have-not

49. Provide, as with some quality

50. Animator Don

51. Fill in the blank with this word: "Cedric ___ of "Little Lord Fauntleroy""

52. Mythical eponym of element #41

54. Fill in the blank with this word: "Albee's "Three ___ Women""

55. Village Voice theater award

56. Popeye's ___ Pea

57. Whirled records?

58. Finish this popular saying: "You are what you_____."

59. Mop & ___ (floor cleaner)

PUZZLE 37

1	2	3	4	■	5	6	7	8	■	9	10	11	12	13
14				■	15				■	16				
17				■	18				■	19				
20			21					22						
23				■	24				■	■	■	■	■	■
■	■	■	25		26		■	27		28	29	30	31	
32	33	34	35		■	36		37		■	38			
39				40		■	41		42					
43			■	44		45		■	46					
47			48		■	49		50			■	■	■	■
■	■	51		52		■	53		54	55	56	57		
58	59	60	61				62							
63				■	64				■	65				
66				■	67				■	68				
69				■	70				■	71				

ACROSS

1. Anna Leonowens, e.g., in "The King and I"

5. Explorer John and others

9. Some soot

14. Z ___ zebra

15. Within: Prefix

16. Fill in the blank with this word: """___ Cassio!": Othello"

17. Alpine sight

18. Mrs. Lincoln's maiden name

19. Fill in the blank with this word: "Aqua ___"

20. Start of a Harvey Keitel jest to a journalist

23. Quiescence

24. Fill in the blank with this word: """If ___ suggest ...""""

25. Have ___ with (know well)

27. ___ note

32. Goose genus

36. Rich, as food

38. Mars: Prefix

39. Dim perception

41. Up for bidding

43. Ziegfeld Follies designer

44. Fill in the blank with this word: "___ prius (trial court)"

46. Ten Commandments word

47. Mall pizza chain

49. Verb-to-noun suffix

51. Fill in the blank with this word: "Feel the ___"

53. One helping with the dishes

58. Diana, with "the"

63. Rice-___

64. Chevrolet model

65. Some voices

66. Times change: in 1990 a statue of this Russian was removed from a Bucharest square after 3 decades there

67. Teen-___

68. Make ___ dash for

69. Sixth-century Chinese dynasty

70. Vintage vehicles

71. Kemo ___ (the Lone Ranger)

DOWN

1. The English translation for the french word: tamoul

2. Uncle Sam's land

3. Speaker of the words in the circled squares, expressed literally

4. Pirouetting, perhaps

5. Certs ingredient

6. Wild Indonesian bovine

7. Swirling

8. This city already had a bad reputation when Lot decided to settle there

9. Words accompanying a smack

10. Take ___ view of

11. From ___ You' (Beatles song)

12. Punnily titled 1952 quiz show "Up to ___"

13. One of the housewives on "Desperate Housewives"

21. Take a piece from

22. Fill in the blank with this word: "___ soul (no one)"

26. Yesterday, in Italy

28. Cheesy snack

29. Young lady of Sp.

30. Trompe l'___

31. Fill in the blank with this word: "___ court (law school exercise)"

32. Fill in the blank with this word: "Dark ___"

33. Workers' rights org.

34. Fill in the blank with this word: "___ spell"

35. Retired, as a prof.

37. Fill in the blank with this word: ""How's it ___?""

40. "More!"

42. What's the ___ trying?'

45. Waterproofing target

48. Kind of class

50. Town shouters

52. The Sun, for example

54. Big tournaments for university teams, informally

55. 1965 march site

56. Cornerstone abbr.

57. Where to sign a credit card, e.g.

58. The English translation for the french word: poîle

59. Riley's "___ Went Mad"

60. The last Mrs. Chaplin

61. Vladimir Nabokov novel

62. Stat starter

PUZZLE 38

1	2	3	4	5	6	7		8	9	10	11		12	13	14	15
16								17					18			
19					20								21			
22				23				24					25			
		26					27				28				29
30	31	32					33				34				
35					36	37	38			39	40			41		
42				43		44			45		46		47		
	48			49						50					
51	52					53					54			55	56
57				58		59		60			61		62		
63			64			65	66				67	68			
69					70				71	72					
	73				74				75				76	77	78
79				80			81					82			
83				84					85						
86				87					88						

ACROSS

1. Delay

8. Icelandic epic

12. Bothers

16. Egg-shaped instrument

17. Black, as la nuit

18. On the qui ___

19. Black-and-yellow beetle

21. Fishing, perhaps

22. "Star Trek" rank: Abbr.

23. Game piece

24. Big ___ Conference

25. Toni Morrison's "___ Baby"

26. The highest ranking manager

30. Bestow

33. Beauty

34. Deck out

35. Gossip

36. Fly high

39. Astern

41. In-flight info, for short

42. Astringent fruit

44. Sundae topper, perhaps

46. Coercion

48. Drained by a river

51. Turn red, perhaps

53. Fill

54. Bend

57. "Give it ___!"

58. Harvest goddess

60. Big pig

62. Apple spray

63. Angry, with "up"

65. ___ el Amarna, Egypt

67. Diner

69. A private detective employe

73. "Act your ___!"

74. "I ___ you one"

75. "Major" animal

76. Trick taker, often

79. Auditory

80. Relating to or befitting Paradise

83. Go for

84. Coastal raptors

85. A degenerate neutron star

86. Song and dance, e.g.

87. Increase, with "up"

88. Permanently attached to a substrate

DOWN

1. Hit the bottle

2. Clickable image

3. Foot pads

4. Victorian, for one

5. Refuse

6. Consecrates with oil

7. Bit of Gothic

architecture

8. Charlotte-to-Raleigh dir.

9. Partially burnt tobacco

10. Catch-22

11. "Gladiator" setting

12. Embodiment

13. Not to your liking

14. Overeat or eat immodestly

15. Caribbean, e.g.

20. Always, in verse

26. Gooey cake

27. Pleasant

28. Character

29. A shag rug

30. Driver's lic. and others

31. "Cool" amount

32. Specializing in diseases

37. "___ Baby Baby" (Linda Ronstadt hit)

38. Kind of dealer

40. Kind of approval

43. Didn't shuffle

45. "I'm ___ you!"

47. Empathize

49. A large edible mushroom

50. Oolong, for one

51. "The War of the Worlds" base

52. Firebrand

55. Propel, in a way

56. Crooked

59. Star of four Hitchcock movies

61. Tenth-anniversary DVD, e.g.

64. Builds

66. Forever, poetically

68. Profits

70. Ninnies

71. Something to chew

72. Klutzy entrances

76. Berry touted as a superfood

77. Sagan of "Cosmos"

78. "... or ___!"

79. Mozart's "L'___ del Cairo"

81. Death on the Nile cause, perhaps

82. Balaam's mount

PUZZLE 39

1	2	3	4		5	6	7	8	9		10	11	12	13
14					15						16			
17					18						19			
20				21			22			23				
		24			25	26			27					
28	29			30					31				32	33
34				35						36				
37			38		39			40	41		42			
43				44			45			46		47		
48					49		50					51		
		52					53				54			
55	56					57				58			59	60
61					62			63	64		65			
66					67						68			
69					70						71			

ACROSS

1. Priestly garb

5. ...

10. Speech problem

14. ...

15. Minimal

16. A chip, maybe

17. "American ___"

18. Job holder?

19. Boris Godunov, for one

20. ...

22. Relating to or extending over

24. "The dog ate my homework," e.g.

27. ¿ la mode

28. Bit of paronomasia

30. "Giovanna d'___" (Verdi opera)

31. Bully

34. ___ Wednesday

35. Like Beethoven

36. Like "The X-Files"

37. Highlander

39. Like a rainbow

42. "To thine own ___ be true"

43. "Encore!"

45. "Your turn"

47. ___ Dee River

48. Difficult or unpleasant

50. "Scream" star Campbell

51. "To ___ is human ..."

52. Binge

53. Go places

55. Crack

58. Small evergreen shrub of Pacific coast of North America

61. Calf-length skirt

62. Force units

65. Sacred Hindu writings

66. "Cast Away" setting

67. Affect

68. Aims

69. Quite a while

70. English exam finale, often

71. Eye affliction

DOWN

1. "___ I care!"

2. Disney dog

3. In a broken-hearted manner

4. A pure form of finely ground silica

5. Priestly garb

6. "Fantasy Island" prop

7. Reading room furniture?

8. Christiania, now

9. Fetor

10. ...

11. A chemical substance that repels insects

12. Antares, for one

13. Make waves

21. Bunch

23. Form of clarified butter

25. Component used in making plastics and fertilizer

26. Cicatrix

28. Heathen

29. Grammar topic

32. Edmonton hockey player

33. Allude

38. Menservants and chauffeurs

40. "... happily ___ after"

41. Vedic mythology

44. "Cool!"

46. Guns

49. One who works hard at boring tasks

54. Overhangs

55. During

56. Galileo's birthplace

57. Bakery selections

59. Email contact info, in slang

60. Beam intensely

63. In-flight info, for short

64. ___ sauce

PUZZLE 40

ACROSS

1. Ball field covering

5. "Check this out!"

9. High-hatter

13. Aroma

14. Blood of the gods

16. Bring on

17. Anniversary, e.g.

18. Nigerian monetary unit

19. ___-friendly

20. Harsh Athenian lawgiver

22. Position in a graded series

24. Cleave

26. Safari sight

27. Kind of first-aid pencil

30. Cousins of crunches

33. Two large muscles of the chest

35. Razor sharpener

37. www.yahoo.com, e.g.

38. Hackneyed

41. "Walking on Thin Ice" singer

42. Ancient Celtic priest

45. Medical exam

48. Overseas

51. Complains

52. Baffled

54. Banquets

55. accident or natural disaster

59. Spoonful, say

62. "God's Little ___"

63. Dostoyevsky novel, with "The"

65. Stronghold captured

66. Synagogue

67. Browning's Ben Ezra, e.g.

68. "I ___ you!"

69. Cozy and comfortable

70. Computer instructions

71. ___ probandi

DOWN

1. Mary in the White House

2. Jewish month

3. Service†club and to promote world peace

4. Maxim

5. A.T.M. need

6. Heroin, slangily

7. Bake, as eggs

8. Ark contents

9. Prevent from entering

10. Not yet final, at law

11. Sundae topper, perhaps

12. European capital

15. Pie cuts, essentially

21. "I'm ___ you!"

23. Aardvark fare

25. Gossip

27. Tater

28. ___ cotta

29. "Wheels"

31. Collection of people or animals or vehicles moving ahead in more†or†less regular formation

32. Navigational aid

34. Back talk

36. Successful runners, for short

39. "___ will be done"

40. Young falcon or hawk

43. With anger

44. ___ any here know me?': King Lear

46. Blue books?

47. Pigment thickly so†that brush

49. Buzzing

50. Fragrant Himalayan tree

53. Accused's need

55. 100-meter, e.g.

56. Bounce back, in a way

57. Jack-in-the-pulpit, e.g.

58. Arcing shots

60. Balsam used in perfumery

61. Aims

64. ___-tac-toe

PUZZLE 41

ACROSS

1. Lover with a ladder, perhaps

7. Modern F/X field

10. Intensifies, with "up"

14. Sportscast feature

15. Ashes holder

16. Bridges of Los Angeles County

17. Written or spoken language

20. At one time, at one time

21. 185-country fiscal agcy.

22. Aquatic mammal

23. "O" in old radio lingo

26. It's a wrap

28. Dash abbr.

31. Rebirth

37. Henry Clay, for one

39. A vessel

40. ___ de force

41. Schuss, e.g.

42. Times to call, in classifieds

43. Antiquated

46. Shaped like a cylinder with tapering ends

48. Like clothes off the rack

50. Alternative to Bowser

51. "Wheels"

52. Minuteman missile

54. Pond buildup

58. "Gimme ___!" (start of an Iowa State cheer)

60. "Cogito ___ sum"

64. Vegetables or flowers

68. Berry touted as a superfood

69. Always, in verse

70. Embodiment

71. Mass number

72. "Don't ___!"

73. Computer key

DOWN

1. Coastal raptor

2. "Laughable Lyrics" writer

3. Chooses, with "for"

4. Cartoon canine

5. Ring bearer, maybe

6. Shaggy Scandinavian rug

7. Coal dust

8. Rude decoration

9. Setting for TV's

"Newhart"

10. Adjoin

11. Cold cuts, e.g.

12. Beep

13. Litigant

18. Bank claim

19. Highland plant

24. "My man!"

25. "___ the fields we go"

27. "It's no ___!"

28. Drive

29. Investigate

30. Trading language

32. Cockeyed

33. Little, e.g.

34. "Well, I ___!"

35. Aegean vacation locale

36. Antique auto

38. Put one's foot down?

41. Storing something

44. Certain École

45. In-flight info, for short

46. Clavell's "___-Pan"

47. "To ___ is human ..."

49. Arabic for "commander"

53. Corrupt

54. Asian nurse

55. Delicate

56. Fat unit

57. Like, with "to"

59. English informant

61. Baptism, for one

62. Buzzing pest

63. Shrek, e.g.

65. Oolong, for one

66. "Dear old" guy

67. "The Three Faces of ___"

PUZZLE 42

ACROSS

1. ___-bodied

5. Enter the picture

11. Amniotic ___

14. Attendee

15. Arrow poison

16. 20-20, e.g.

17. A natural or surgical joining of parts

19. www.yahoo.com, e.g.

20. Student's worry

21. Fit to be taken in

23. Absorbed, as a cost

24. Again

26. ...

27. The New Yorker cartoonist Edward

29. "___ we forget"

32. Boost

33. ___ gestae

35. 10 C-notes

37. In-flight info, for short

38. A difficult problem

41. Trick taker, often

43. Chimney channel

44. ...

45. Light

47. Cookbook abbr.

49. Philippine banana tree

53. Beethoven's "Archduke ___"

54. Old Chinese money

56. Delivery person?

57. Paste of olive oil, chilli, garlic and spices

61. Easily moved

63. ___-Wan Kenobi

64. Executive perk

66. Amscrayed

67. Bony

68. Prefix with phone

69. Holiday drink

70. Someone's foot

71. "Aeneid" figure

DOWN

1. Small terrestrial lizard of warm regions of the Old World

2. Kind of shark

3. Take-charge sort

4. At one time, at one time

5. ___ squash

6. Beat

7. Ace

8. "___ on Down the Road"

9. "Mi chiamano Mimi," e.g.

10. Medical advice, often

11. Short and blunt

12. Emergency delivery, of a sort

13. Keyboard instrument

18. Bit that's read in phyllomancy

22. Afflict

25. Covered or soaked with a liquid

28. "To ___ is human ..."

30. "___ Cried" (1962 hit)

31. Contemptible one

34. Bank deposit

36. By and large

38. Make soiled, filthy, or dirty

39. Center

40. "Chicago" lyricist

41. An instrument of the saxhorn family

42. Water buffalo of the Philippines

46. Hawaiian dish

48. Olive Garden selection

50. Spy novelist Eric

51. Arboreal nocturnal mammal

52. Microscope slide bunch

55. Before the due date

58. Sean Connery, for one

59. "Eh"

60. Mercury, for one

62. #1 spot

65. After expenses

PUZZLE 43

1	2	3	4	5		6	7	8	9		10	11	12	13
14						15					16			
17				18						19				
20				21						22				
		23					24	25						
26	27	28				29					30	31	32	
33					34				35					
36				37	38					39				
40		41		42					43	44				
45			46			47	48							
		49				50								
51	52	53				54					55	56	57	
58				59	60				61					
62				63				64						
65				66				67						

ACROSS

1. Beth's preceder

6. Door fastener

10. Hollow pastry

14. Beau

15. Auspices

16. Arch type

17. Subject to trial by court-martial

20. Harvest goddess

21. Hindu divinity

22. Salome's septet

23. Bang-up

24. Limerick, e.g.

26. Bride is pregnant

33. Places for pews

34. "___ Ng" (They Might Be Giants song)

35. Blacken

36. ___ grecque (cooked in olive oil, lemon juice, wine, and herbs, and served cold)

37. Nocturnal badger-like carnivore

39. Egg cells

40. Pink, as a steak

42. Parenthesis, essentially

43. "Unforgettable" singers

45. Get very angry and fly into a rage

49. Gobs

50. Art subject

51. Flight segment

54. Drop

55. ___ cross

58. Occurring in the same period of time

62. Asian nurse

63. Qualm

64. Congo's old name

65. Hair goops

66. Civil War side, with "the"

67. Book of maps

DOWN

1. "Giovanna d'___" (Verdi opera)

2. Aerial maneuver

3. Flightless flock

4. The "p" in m.p.g.

5. Someone who performs dangerous stunts

6. Bliss

7. City on the Yamuna River

8. Be in session

9. Telekinesis, e.g.

10. Forceful verbal attack

11. ___ fruit

12. Experience

13. Revenuers

18. Children's ___

19. Affirm

23. Absorbed, as a cost

24. French door part

25. Face-to-face exam

26. Devour greedily

27. Acceptable by Muslim law

28. Egg producer

29. Court session

30. "Get ___ of yourself!"

31. Bench banger

32. Clear, as a disk

37. Flat floater

38. Song and dance, e.g.

41. ...

43. A brilliant solo passage

44. "___ moment"

46. "Watch out!"

47. Liveliness

48. ___ Bowl

51. Heroin, slangily

52. Heavy reading

53. Of or related to the anus

54. Lady of Lisbon

55. Drudgery

56. Halo, e.g.

57. A Swiss army knife has lots of them

59. Automobile sticker fig.

60. When it's broken, that's good

61. "Dig in!"

PUZZLE 44

1	2	3	4		5	6	7			8	9	10	11	12
13					14			15		16				
17					18					19				
20				21					22					
23						24					25	26	27	
		28		29	30				31	32				
33	34	35		36				37						
38			39	40			41							
42						43				44				
45					46				47	48				
49				50					51		52	53	54	
		55	56	57			58	59						
60	61				62					63				
64					65					66				
67					68					69				

ACROSS

1. Heroin, slangily

5. ___ Wednesday

8. Beauty pageant wear

13. Astronaut's insignia

14. A crybaby

16. Knight's "suit"

17. A chip, maybe

18. Way, way off

19. Three-masted vessel

20. Illusory state of wellbeing

23. Tear open

24. Directly

25. Biochemistry abbr.

28. Seeks power or success

33. CD follower

36. A hand

37. Ancient debarkation point

38. Inclined toward or displaying love

41. The highest level or degree attainable

42. Garam ___ (Indian spice mixture)

43. Elmer, to Bugs

44. Diffident

45. Someone who takes photographs professionally

49. Amniotic ___

50. Dusk, to Donne

51. Demands

55. Tree having palmate leaves and large clusters

60. Accused's need

62. Christian name

63. Doing nothing

64. "Belling the Cat" author

65. Make out, in Middlesex

66. Gift on "The Bachelor"

67. Attached to the collar of a draft horse

68. Mail place: Abbr.

69. The America's Cup trophy, e.g.

DOWN

1. Big mess

2. Religious law

3. Mary of "The Maltese Falcon"

4. Manx, e.g.

5. "By yesterday!"

6. Chesterfield, e.g.

7. Acted the miser

8. Checker, perhaps

9. ...

10. Waxy substance

11. ___ v. Wade

12. Parenthesis, essentially

15. Dog's coat?

21. Marienbad, for one

22. Corrupt morally

26. Poet's "below"

27. Dilettantish

29. A destructive action

30. Pandowdy, e.g.

31. Branch

32. Cheesecake ingredient?

33. Level connectors

34. 1935 Triple Crown winner

35. ...

39. Sylvester, to Tweety

40. 1969 Peace Prize grp.

41. Absorb, with "up"

43. Ballroom activity

46. Marsh growth

47. "Star Trek" rank: Abbr.

48. Stop working

52. Back, in a way

53. Edible seaweed

54. Animal in a roundup

56. "O" in old radio lingo

57. Saws with the grain

58. Jazz score

59. "Cogito ___ sum"

60. "I see!"

61. Grassland

PUZZLE 45

ACROSS

1. Teeth

6. Garbage

11. Made of, containing or resembling wood

13. Track orb

15. W.W. I battle locale

16. Made of, containing or resembling wood

17. A grey or greenish-blue mineral consisting of aluminum silicate.

18. Fighting

19. "The Snowy Day" author ___ Jack Keats

20. Flashed signs

22. Frosts, as a cake

23. Dapper

25. 100 cents

26. Bungle, with "up"

27. "I ___ return"

29. A murderer who slashes the victims with a knife

31. Otalgia

33. A metric†unit of weight equal†to one

thousandth of a kilogram.

36. Aden's land

40. Face-to-face exam

41. Wading†birds of warm regions having long slender down-curved bills...

43. Increase

45. Hamster's home

46. Indian bread

47. ___ mortals

48. Beseech

50. Titanic

53. Perturb

54. Tip

55. Holmes and Moriarty, e.g.

56. Spring times

57. Give an interpretation or explanation to

58. Alibi

DOWN

1. An inflammation of the mucous†membrane lining the nose

2. English artist noted for a series of engravings

3. Dragonflies and damselflies

4. Peewee

5. Confined, with "up"

6. A system of belief based on mystical insight into the nature of God and the soul

7. Engine parts

8. Simultaneously

9. A woodworker who joins pieces of wood with a splice

10. Of something totally lacking in saturation and therefore having no hue

11. Come to

12. "The Canterbury Tales" pilgrim

13. Having the shape of a sphere or ball

14. Amount of hair

21. Home to nearly 70% of all people

24. Primordial matter

28. Create laminate by bonding sheets of material with a bonding material

30. Equal

32. A refund of some fraction of the amount paid

33. A person who operates a farm

34. When your clothes wear out?

35. Sent signals to

37. Token

38. The rider of a horse used in eventing

39. The baby's room

40. Arctic ___

42. Marsh bird

44. Equals

49. "___ on Down the Road"

51. Cousins of the ostrich

52. Atomizer output

PUZZLE 46

ACROSS

1. Branch

4. An aromatic exudate from the mastic tree

11. Fergie, formally

16. Morgue, for one

17. Snakes

18. Chip away at

19. Habitual

21. Crosses

22. Bridges of Los Angeles County

23. Elmer, to Bugs

24. Amiens is its capital

25. The chief mountain range of western North America

31. An orange isomer

34. "Come to think of it ..."

35. Tokyo, formerly

36. Trick taker, often

37. 20-20, e.g.

38. Give birth

41. A broad cartridge belt

44. Affranchise

45. Locale

46. Cake part

48. "The Last of the Mohicans" girl

52. Head, for short

55. A closed litter

58. First to think of or make

62. Conk out

63. Engine speed, for short

64. Popular fish for show

65. Blue

66. Tusked mammals

68. Climbing over, crawling through,

72. Cupid's boss

73. Bull markets

74. "___ of Eden"

78. Burgundy grape

79. A curved section or tier

83. Anoint, old style

84. Eat at a restaurant or at somebody else's home

85. "Spy vs. Spy" magazine

86. Stage item

87. Hard-to-call contests

88. Dash lengths

DOWN

1. Kuwaiti, e.g.

2. Bumpkin

3. ___ Verde National Park

4. Back-to-work time: Abbr.

5. Appropriate

6. "___ Cried" (1962 hit)

7. 60s coloring method

8. Going to the dogs, e.g.

9. Kid's name

10. "Casablanca" pianist

11. Microorganisms or viruses

12. Bouquet

13. Having ample space

14. Extras

15. "For ___ a jolly ..."

20. 100 cents

24. Tangle

26. "Beetle Bailey" dog

27. Provide with an overhead surface

28. Show respect, in a way

29. "It's no ___!"

30. Costa del ___

31. Street fleet

32. Healthy berry

33. Cost of living?

38. One with a crowbar, perhaps (Var.)

39. Brought into play

40. Carve in stone

42. Kosher ___

43. "___ on Down the Road"

47. "M*A*S*H" role

49. Boat propellers

50. Opportune

51. Kind of dealer

53. Excited in anticipation

54. Kind of concerto

56. Lagerl^f's "The Wonderful Adventures of ___"

57. Part of a bird's beak

58. Alias

59. "Semiramide" composer

60. Infusion of e.g.

61. 100 lbs.

66. Weak and ineffectual

67. ___-friendly

69. American chameleon

70. Antique shop item

71. Some tournaments

75. #1 spot

76. Brickbat

77. "Bill & ___ Excellent Adventure"

78. When it's broken, that's good

79. "Silent Spring" subject

80. Former French coin

81. Trophy

82. "___ alive!"

PUZZLE 47

ACROSS

1. Falling flakes

5. Commoner

9. Bridal path

14. Novice

15. "Where the heart is"

16. Fat

17. Biblical birthright seller

18. "Not on ___!" ("No way!")

19. Allotment

20. Means of support

22. Big Indian

24. Dusk, to Donne

25. In a dim indistinct manner

27. Decorated, as a cake

29. Watergate, e.g.

32. Bulbous herb of southern Europe

35. Egyptian Christian

36. Sees

39. Hokkaido native

40. Increase, with "up"

41. 1935 Triple Crown winner

42. Fold, spindle or mutilate

43. Page

45. Small tree native to the eastern United States

46. Durable wood

47. A sculpture representing a human or animal

49. Wore

51. ...

52. Purple shade

53. "Bingo!"

55. ___ Strip

57. Rubenesque

61. Black

63. Court attention-getter

65. ___-bodied

66. ___ wrench

67. Elliptical

68. Poker action

69. Fold

70. ___ mortals

71. At one time, at one time

DOWN

1. Increase, with "up"

2. Not yet final, at law

3. Face-to-face exam

4. Would not

5. a book containing a compilation of pharmaceutical

6. Court ploy

7. File

8. ___ carotene

9. Petting zoo animal

10. "Rocky ___"

11. Top speed

12. Vermin

13. "Our Time in ___" (10,000 Maniacs album)

21. A hand

23. A puzzle

26. Andean animal

28. Victorian, for one

29. Picket line crossers

30. Cleanser brand

31. Capable of being pacified

33. Absurd

34. Fixed

37. Drink from a dish

38. ___ apso (dog)

44. "Harper Valley ___"

46. Introduce

48. Pressing

50. Egg cells

52. May have

53. "By yesterday!"

54. Campus building

56. PBS show "by kids, for kids"

58. Alpine transport

59. Misfortunes

60. Cabbage

62. Grassland

64. Ring bearer, maybe

PUZZLE 48

1	2	3	4		5	6	7	8	9		10	11	12	13	14	15
16					17						18					
19				20							21					
22						23					24					
25				26				27	28							
29				30				31					32	33	34	
35				36			37				38					
		39			40					41						
	42	43		44			45	46								
47				48			49									
50			51	52				53			54	55	56			
57				58				59			60					
		61			62			63								
64	65	66			67			68								
69					70			71								
72					73					74						
75					76					77						

ACROSS

1. Priestly garb

5. Breaks

10. ___ Tuesday

16. One of TV's Simpsons

17. Accessory

18. Doctor

19. Record of scores

21. Sleeve type

22. Curry spice

23. ___ grecque (cooked in olive oil, lemon juice, wine, and herbs, and served cold)

24. High marks

25. Unsatisfactoriness

29. Fill

30. "It's no ___!"

31. Noticeably masculine

32. Egg cells

35. "... ___ he drove out of sight"

36. Dracula, at times

37. A whetstone for use with oil

39. Comedian Bill, informally

40. A chorus line

41. Disorder

42. Following the ideas of Marx

47. Greek earth goddess: Var.

48. Amount of work

49. Cashew, e.g.

50. A follower who carries

out

53. Stop on a crawl

54. Amniotic ___

57. Cracker Jack bonus

58. Auto parts giant

59. Fold, spindle or mutilate

60. Pie perch

61. An educated and intellectual elite

64. Rupture

67. "Rocks"

68. Like Cheerios

69. A whole lot

70. Spreading out in different directions

72. California county

73. A unit of subjective loudness

74. "Trick" joint

75. Least tainted

76. "Fiddler on the Roof" role

77. Lentil, e.g.

DOWN

1. Shed flowers and leaves

2. Pertaining to, or having, lacunae

3. React with bromine

4. Big step

5. ___ gestae

6. Wear out completely

7. An ancient upright stone

8. Milk-Bone biscuit, e.g.

9. Cal. col.

10. Screams

11. Football coach's equipment

12. In things

13. ___ podrida

14. Butcher's offering

15. Coastal raptors

20. Charlotte-to-Raleigh dir.

26. Seeming

27. Served Time?

28. Hose material

32. ___ probandi

33. Clothe

34. Balaam's mount

36. Exchange blows

37. Spicy stews

38. Bypass

39. Jam

40. Discordant

42. ___ Clinic

43. Breezy

44. Inclination

45. Accustom

46. Center

47. Clock standard: Abbr.

51. Nine of diamonds?

52. Computer program input

53. Show

54. "The ___ Madonna" (Raphael painting)

55. Property recipient, at law

56. Made a cacophonous sound

59. Thousandth of a millimeter

60. Pursues

61. Absurd

62. Flexible

63. Holiday drink

64. Jazz score

65. Almond

66. Back

70. 007, for one

71. Fed. construction overseer

PUZZLE 49

ACROSS

1. Baker's abbr.

5. Healthy berry

9. Prove useful

14. Southwestern pot

15. Group standard

16. From that time

17. Like some tickets

19. The male organ of copulation.

20. There's no accounting for it

21. Freshman cadet

23. 'C'___ la vie!"

24. Suffix of some ordinal numbers

25. Main computer part (Abbr.)

27. Perfect scores

29. Having a momentous impact

31. Basil or sage, e.g.

35. The male organ of copulation.

38. Of a previous time

39. Hawaii island

40. The male organ of copulation.

43. Jack-o'-lantern feature, perhaps

44. 'Fargo' affirmative

45. 'Amen!"

46. It may be due, get the point?

47. Up for grabs

49. Remit

51. Turkish title

52. 'The Travels of Marco Polo" creature

55. Cell substance

58. Do Circe's job

60. Brenda the reporter

62. The male organ of copulation.

64. In a crude and unskilled manner.

66. Passerine bird

67. Farm storage building

68. Summit position

69. Subdivision maps

70. Diving bird

71. Speak wildly

DOWN

1. Rich cake

2. Puff up

3. Some winter wetness

4. Suck wind

5. Tiny colonist

6. Capital as contrasted with the income derived from it.

7. Seed coating

8. A very contagious infection of the skin; common in children.

9. Tree with fluttering leaves

10. Be in the running

11. First name of Henry VIII's second

12. Tries to reduce swelling, in a way

13. In the event

18. Chicanery

22. According with custom or propriety.

26. Trifling

28. Would's cousin

29. Restricts in amount

30. Like Rambo

32. Victuals

33. Ostrich cousin

34. Repress in memory

35. Advantage

36. Sorvino of ''Mighty Aphrodite"

37. Sponsorship

41. A large continuous extent of land.

42. Gratify

48. Turkey dangler

50. Distinctive spirit of a culture

52. Certain dreadlocked Jamaican, briefly

53. Lightweight synthetic fabric

54. Underground chamber

55. 'Please reply"

56. A short or waste piece or knot of wool.

57. Invisible emanation

59. Proper partner

61. Peter the Great's title

63. Retrieve

65. Needing replenishment

PUZZLE 50

1	2	3	4	5		6	7	8	9	10		11	12	13
14						15						16		
17				18								19		
20					21					22	23			
		24	25				26	27						
28	29	30					31							
32						33					34	35	36	
37				38	39					40				
41			42	43					44	45				
		46					47	48						
49	50	51				52								
53						54				55	56	57	58	
59			60	61	62				63					
64			65					66						
67			68					69						

ACROSS

1. CaffË ___

6. Apprehension

11. Fed. construction overseer

14. ___ squash

15. Like "The X-Files"

16. Amscrayed

17. An English verb

19. "A jealous mistress": Emerson

20. Hammer part

21. Born, in bios

22. Indian coin

24. Ten-armed oval-body

28. Biker's hot-dog maneuver

31. Length x width, for a rectangle

32. Verb with thou

33. A rude decoration

37. ___ be an honor'

38. Kind of room

40. Calendar abbr.

41. Adjust anew

44. Barely speak

46. Christmas season

47. By and large

49. A preview to test audience reactions

53. Crush

54. Cashew, e.g.

55. Shot in the arm

59. Ashes holder

60. Fight with Oscar winner Sally?

64. ì___ boom bah!î

65. "La BohËme," e.g.

66. Barely beat

67. Pillbox, e.g.

68. Roentgen's discovery

69. CÈzanne contemporary

DOWN

1. Arctic native

2. Advil target

3. Bolted

4. A psychological state

5. "Star Trek" rank: Abbr.

6. Computer key

7. Celebrate

8. "... ___ he drove out of sight"

9. Balloon filler

10. Debriefing and make him report

11. Result of some plotting

12. Eastern wrap

13. Chips in

18. The "A" of ABM

23. Org. that uses the slogan 'Aim High'

25. Final: Abbr.

26. All ___

27. Phi Delt, e.g.

28. Blender sound

29. "Unimaginable as ___ in Heav'n": Milton

30. Icelandic epic

33. "___ lost!"

34. "American ___"

35. Blue hue

36. Gross

38. Mouthful

39. ___-friendly

42. Pair

43. Singles player

44. Nod, maybe

45. Series on which Clint Eastwood played Rowdy Yates

47. Throat hangers

48. Locale

49. "Pipe down!"

50. Water wheel

51. Artist Max

52. Admittance

56. Burglar

57. "Guilty," e.g.

58. "What are the ___?"

61. 30-day mo.

62. Oolong, for one

63. Disobeyed a zoo sign?

Solutions

Puzzle Solution 1

```
K U N G ■ B S A W ■ A X I O N
I N I T ■ A L L E ■ L A N A I
S E A O F L O V E ■ T N U T S
S A C ■ I L A Y ■ C A A N ■
E S I A S O N ■ P H I D I A S
S Y N C H ■ M A I ■ U S C G
■ H Y P H E N S ■ O R T
C A P E ■ A E R E O ■ O N O S
H M O ■ M D C L X V I
U B E R ■ P D S ■ A L E R O
G O T O S E A ■ L E N S C A P
■ I N T R ■ M E A N ■ H C E
L O C A L ■ B R I G A D O O N
A T A L E ■ I O N E ■ L E O I
S O L D O ■ S N E R ■ I S N T
```

Puzzle Solution 2

```
N Y A D ■ E N A M ■ D E C I R
G A L A ■ N O T A ■ I N L O W
O R I S ■ U P R I ■ T L E T T
R E D H E R R I N G S ■ R A J
■ O E O ■ M A Y I ■
H O T E L ■ B R E L ■ M O C S
Y O U D I G ■ I N E ■ E R A S
M O N I C A M O U N T A I N S
A L I T ■ M E T ■ A N N O Y S
N A S O ■ E R S T ■ A T N O S
■ R O T C ■ A P I ■
B A S ■ N E U T R A L Z O N E
S T A L E ■ T O S S ■ A N O A
M O D A L ■ I M U S ■ C E N S
T R A P S ■ O M S K ■ H A S T
```

Puzzle Solution 3

D	I	E	T	E	T	I	C	■	T	H	E	C	A	R
I	D	L	O	V	E	T	O	■	W	O	R	T	H	Y
S	O	F	T	E	N	E	R	■	I	L	O	N	A	S
■	■	E	N	T	R	E	A	T	Y	■	■	■	■	■
A	S	O	R	T	■	■	■	M	C	G	U	I	R	E
P	H	L	M	■	B	W	A	S	H	■	L	N	E	S
R	E	O	■	H	O	L	T	■	■	S	N	A	I	L
■	B	R	I	T	N	E	Y	S	P	E	A	R	S	■
P	A	O	L	O	■	■	O	D	I	U	■	M	S	F
E	N	S	E	■	T	R	U	S	T	■	Y	O	U	D
I	G	O	T	C	H	A	■	■	B	E	R	E	A	■
■	■	■	H	A	B	A	N	E	R	A	■	■	■	■
S	L	O	C	U	M	■	R	E	T	O	R	T	E	D
S	O	R	A	R	E	■	E	A	T	A	L	O	N	E
S	N	A	R	L	S	■	S	T	U	D	Y	F	O	R

Puzzle Solution 4

A	T	C	O	■	L	E	N	O	■	I	T	A	K	E
R	I	T	Z	■	I	C	E	R	■	D	O	R	A	L
S	E	N	T	E	N	C	E	D	T	O	J	A	I	L
E	S	S	■	R	E	E	D	I	T	■	O	I	S	E
■	■	■	I	A	M	■	S	E	I	S	■	S	E	N
E	R	E	C	T	E	D	■	■	M	E	I	E	R	■
L	O	S	E	O	N	E	S	H	E	A	D	■	■	■
Y	Z	E	J	■	■	S	K	I	■	■	E	R	I	N
■	■	■	A	I	R	C	A	N	A	D	A	D	R	Y
■	B	A	M	B	I	■	■	T	H	E	L	A	S	T
B	A	N	■	■	A	C	R	O	■	E	N	S	■	■
O	R	G	S	■	■	C	O	M	E	A	T	■	S	S	W
S	T	L	O	U	I	S	O	R	D	I	N	A	L	S
S	A	I	N	T	■	■	A	R	L	O	■	O	N	Y	M
A	B	A	S	E	■	■	R	E	E	F	■	N	E	S	S

Puzzle Solution 5

M	A	C	H	U	■	T	A	L	C	■	C	E	R	A
E	L	L	I	S	■	O	M	A	R	■	E	L	A	N
S	P	O	T	L	E	S	S	S	P	O	T	L	E	S	S
O	H	T	O	■	U	S	O	■	T	E	E	V	E	E
■	■	R	Z	B	■	■	M	O	E	S	■			
U	S	B	■	T	I	M	E	A	N	D	T	I	D	E
N	A	A	■	E	E	L	E	D	■	A	T	I	S	
J	C	R	E	W	■	C	M	A	■	S	S	S	S	S
A	R	A	N	■	A	C	M	E	S	■	A	C	A	
M	A	K	E	C	E	R	T	A	I	N	■	T	O	I
■	R	O	A	R	■	■	N	S	C	■				
H	A	N	G	E	R	■	I	L	E	■	L	B	O	S
S	T	A	I	N	L	E	S	S	S	T	E	E	L	S
E	L	I	Z	■	Y	A	N	A	■	O	F	A	G	E
T	I	L	E	■	B	L	O	T	■	S	T	R	A	T

Puzzle Solution 6

A	D	L	E	R	■	G	R	A	D	E	■	O	W	S
R	I	O	D	E	■	N	A	U	R	U	■	T	A	H
M	E	T	A	L	W	O	R	K	E	R	■	E	Y	E
E	S	T	■	A	I	M	E	■	E	N	R	O	L	
R	E	A	D	Y	F	O	R	T	A	K	E	O	F	F
■	I	S	I	N	■	T	R	A	C	■				
K	H	B	D	■	Z	O	I	C	■	M	M	I		
C	O	L	O	R	A	D	O	P	L	A	T	E	A	U
S	L	Y	■	A	L	M	A	■	A	H	E	M		
■	O	T	E	A	■	D	I	A	L	■				
J	U	M	P	I	N	J	A	C	K	F	L	A	S	H
T	H	I	E	F	■	C	A	O	R	■	P	L	A	
I	A	N	■	I	N	C	A	R	N	A	T	I	O	N
L	U	I	■	E	A	U	D	E	■	M	I	N	S	K
E	L	M	■	S	E	T	S	A	■	E	S	T	H	S

Puzzle Solution 7

C	Y	T	O	■	B	O	F	F	O	■	H	O	K	E
N	A	I	F	■	C	A	R	O	M	■	E	T	A	T
B	Y	E	S	■	E	X	U	R	B	■	A	T	R	A
C	A	R	T	I	L	A	G	E	■	S	R	O	O	T
■	■	■	L	L	C	■	P	I	S	T	I	L	S	
M	E	R	G	E	■	A	G	A	S	S	I	■		
T	I	E	R	R	A	■	O	W	S	■	N	C	O	S
G	R	U	E	■	D	B	A	S	E	■	T	O	R	E
S	E	S	E	■	A	O	L	■	I	T	H	I	N	K
■	N	E	T	T	O	N	■	A	E	R	E	O		
C	I	R	C	L	E	S	■	I	B	N	■			
A	S	S	A	I	■	W	I	N	E	T	H	I	E	F
L	A	I	R	■	S	A	C	E	R	■	I	M	N	O
M	A	D	D	■	O	N	E	P	M	■	C	A	L	X
A	C	E	S	■	W	A	R	M	S	■	S	M	S	H

Puzzle Solution 8

G	R	P	S	■	N	O	M	E	■	S	R	O	O	F
E	I	L	A	■	E	V	A	N	■	P	E	A	L	E
N	E	A	P	■	H	O	R	N	R	I	M	M	E	D
O	N	Y	O	U	■	Y	E	A	R	N	■			
A	Z	E	R	B	A	I	J	A	N	I	■	L	E	S
S	I	R	■	O	M	N	■	A	T	B	E	S	T	
■	I	L	E	A	V	E	■	A	T	T	U			
E	L	E	C	T	R	I	C	A	L	S	T	O	R	M
C	E	R	E	■	D	R	R	U	T	H	■			
C	H	E	Q	U	E	■	T	B	A	■	R	M	N	
E	I	S	■	G	E	T	T	H	E	P	B	O	U	T
■	A	L	O	U	S	■	H	A	L	F	E			
S	T	A	T	I	C	L	I	N	E	■	R	E	F	S
C	A	V	A	E	■	I	D	E	D	■	N	O	I	T
M	O	O	N	S	■	P	E	K	E	■	S	S	N	S

Puzzle Solution 9

```
S P C A ■ A I L S ■ A N D I M
T A L L ■ H A U T ■ R E E S E
B I O U ■ O N C E ■ C A L M A
D R Y M A R T I N I S ■ S S T
■ ■ I S S H E ■ T I T ■
S T A ■ Y E O ■ E N H A S E
L E N D L ■ R A H M ■ O L E O
A N I M A L P R E S E R V E S
E T T A ■ B E N S ■ R A I M I
D H A R M A ■ I S I ■ N E N
■ K O R ■ I T H C A ■
F L Y ■ I S T H A T A F A C T
Q U E E R ■ R O T E ■ O D E A
H A S T E ■ A P E T ■ O U R N
P U M A S ■ M E R L ■ L E S T
```

Puzzle Solution 10

```
C R A M P ■ E S S I ■ S S S S
P E R O U ■ D T E N ■ T O A T
A G I R L ■ T I R E ■ A R I Z
S T E E L ■ V C R S ■ R A N A
■ A T P ■ K A S E M ■
S A H U A R O S ■ E X A L T S
I R E ■ B E K A A ■ T P I K E
X M A S ■ P O N C A ■ S E L L
T E R P S ■ K D C F E ■ T E L
H E N R Y V ■ S T O P L O S S
■ A D A M T ■ R I S ■
D E L I ■ C A O R ■ G E T A T
A M E N ■ A C N E ■ O V E R S
R E C E ■ N A E S ■ N E A T O
E S T D ■ T U S H ■ E N T E S
```

Puzzle Solution 11

A	B	A	C	A		A	B	E	L		R	U	D	D
G	E	N	U	S		N	E	R	O		E	V	E	R
O	A	T	E	S		T	R	I	O		D	E	L	E
G	U	I	D	E	L	I	N	E	S		C	A	V	A
		T	I	P	I		E	R	A	S	E	R		
M	I	S	S		L	A	N		N	U	B			
I	C	A	O		A	S	I	A		I	B	E	A	M
N	E	G	L	E	C	T		F	I	N	A	N	C	E
I	S	A	A	C		O	D	O	R		G	O	E	R
		R	H	O		E	R	A		E	L	S	E	
M	A	Y	P	O	P		B	E	T	A				
A	N	O	A		T	E	A	S	E	R	V	I	C	E
N	O	U	N		O	R	C	A		R	A	D	O	N
I	D	L	E		U	G	L	I		A	L	O	U	D
C	E	L	L		T	O	E	D		Y	E	L	P	S

Puzzle Solution 12

U	R	I	C		O	B	O	E		P	A	P	A	W
V	E	D	A		R	A	P	S		A	L	I	B	I
E	V	E	N		G	A	T	T		P	L	E	A	T
A	U	S	T	R	A	L	I	A	D	A	Y			
	E	T	H	O	S		O	T	I	C		S	H	E
	I	M	M	U	N	E	S	Y	S	T	E	M		
W	E	E		A	I	L	S			A	O	N	E	
A	D	J	U	N	C	T		O	B	S	C	E	N	E
H	I	E	S			I	C	A	O		P	A	R	
O	C	C	A	S	I	O	N	A	L	L	Y			
O	T	T		A	R	C	S		L	O	E	S	S	
		O	N	E	U	P	M	A	N	S	H	I	P	
A	C	I	N	I		L	I	E	D		M	A	N	E
S	P	O	U	T		U	R	G	E		A	L	G	A
P	A	N	S	Y		S	E	A	S		N	E	S	T

Puzzle Solution 13

R	U	T	S		S	H	U	T	S		M	A	K	O
A	B	U	T		P	E	R	I	L		I	C	E	D
M	O	T	E		A	M	E	B	A		S	T	Y	E
P	A	T	R	O	N	S	A	I	N	T	S			
S	T	I	N	G			A	G	A	I	N	S	T	
		A	L	K	I	E		L	L	A	N	O		
O	T	T		E	N	T	R	A	N	C	E	W	A	Y
C	A	R	S		E	E	R	I	E		S	A	F	E
T	W	E	L	V	E	M	O	N	T	H		B	U	D
A	S	S	A	I			R	U	S	E	S			
D	E	S	P	O	T	S				R	I	D	E	S
		B	L	A	T	H	E	R	S	K	I	T	E	
C	O	L	A		F	R	E	Y	A		K	O	H	L
O	W	E	N		F	I	R	E	D		I	D	O	L
P	L	U	G		Y	A	R	D	S		M	E	S	S

Puzzle Solution 14

S	A	D	H	U		L	A	P	P	I	S	H		G	O	T
A	L	I	E	N		O	C	E	A	N	I	A		E	M	U
E	L	E	P	H	A	N	T	G	R	A	S	S		N	E	T
			O	N	E		L	I	P		G	I	G	O		
N	E	P	A	L	I		Y	E	S	T	E	R	Y	E	A	R
B	E	L	L	Y	L	A	U	G	H		M	O	P			
C	L	O	G		E	L	K		P	O	D		S	H	E	
	D	A	M		M	O	O		A	T	E		K	E	G	
V	I	D	E	O	C	O	N	F	E	R	E	N	C	I	N	G
O	R	E		J	U	S		T	I	C		T	A	D		
W	E	D		A	R	T		D	E	B		S	L	I	T	
		I	V	E		R	E	E	L	E	C	T	I	O	N	
F	O	X	T	E	R	R	I	E	R		N	E	E	D	N	T
A	V	E	S		O	A	R		R	E	M					
R	U	B		B	E	L	L	I	C	O	S	E	N	E	S	S
A	L	E		U	R	E	T	E	R	S		N	E	W	E	L
D	E	C		R	E	S	O	R	T	S		T	E	E	N	Y

Puzzle Solution 15

O	M	A	N		S	T	R	A	P	S		A	T	S	E	A
B	O	N	E		T	O	I	L	E	T		G	O	W	N	S
I	N	D	E	T	E	R	M	I	N	A	T	E	N	E	S	S
	S	O	D	O	M				N	T	H		G	E	N	E
S	T	R	E	P		P	U	B		A	A	H		P	A	S
T	E	R	R	I	T	O	R	I	A	L	W	A	T	E	R	S
E	R	A	S		H	E	L	P	S			M	E	R	L	E
P	A	N		P	A	T		A	T	E		M	A	S	S	E
		H	A	I	R	T	R	I	G	G	E	R				
C	A	M	E	L		Y	E	T		G	A	D		R	P	M
A	C	A	R	I			R	I	G	E	L		S	E	R	E
S	T	R	O	N	G	I	N	T	E	R	A	C	T	I	O	N
T	I	S		G	A	M		E	M	S		R	A	N	T	S
A	N	U	S		S	P	A				C	O	P	S	E	
W	I	P	E	T	H	E	S	L	A	T	E	C	L	E	A	N
A	D	I	E	U		D	E	I	C	E	S		E	R	S	E
Y	E	A	R	N		E	A	S	E	L	S		S	T	E	W

Puzzle Solution 16

S	E	T	A		S	C	A	D			B	O	R	R	O	W	S
E	T	U	D	E		O	G	R	E		M	A	U	N	A	L	O	A
C	H	R	O	M	A	T	I	C	A	B	E	R	R	A	T	I	O	N
		G	R	A	S	S		R	O	T	E			T	O	L	D
I	R	E	N	I	C		P	I	T	A		U	S	A			
D	A	N		L	I	F	E	T	H	R	E	A	T	E	N	I	N	G
O	N	E	S		A	C	E		T	R	E	E		M	O	A
L	I	V	E	R	Y	C	O	M	P	A	N	Y		D	R	A	W	S
		T	O	O	T	S		O	P	A	L	S		E	G	I	S
C	H	A	S	E	R		A	L	P		L	E	M	O	N	Y
H	O	M	O		E	M	P	T	Y		T	A	I	G	A
A	L	O	N	G		E	L	E	P	H	A	N	T	G	R	A	S	S
F	L	U		N	I	N	A		A	C	T		K	N	E	E
F	O	R	W	A	R	D	T	H	I	N	K	I	N	G		T	A	T
		O	W	E		O	R	G	Y		O	U	S	E	L	S
A	T	O	M		T	O	P	I		A	V	I	A	N
G	A	R	B	A	G	E	D	I	S	P	O	S	A	L	U	N	I	T
E	M	B	A	L	M	E	D		E	A	C	H		E	C	A	S	H
S	E	S	T	E	T	S		D	R	A	Y		E	L	M	Y

Puzzle Solution 17

D	E	L	L	S	■	T	R	U	S	S	■	C	H	A
E	M	A	I	L	■	R	A	T	T	Y	■	H	E	S
C	O	M	B	I	N	A	T	I	O	N	L	O	C	K
A	T	E	■	T	I	C	■	L	A	C	E	R	T	A
M	I	L	L	■	C	E	D	I	■	H	A	D	O	N
P	O	L	Y	M	E	R	I	Z	E	S	■	A	R	C
■	N	A	S	A	■	■	G	E	M	■	E	L	S	E
■	■	■	E	R	G	O	■	D	U	E	T	■	■	■
T	E	N	S	■	O	I	L	■	■	T	H	R	U	■
R	N	A	■	R	O	L	E	P	L	A	Y	I	N	G
A	V	I	S	O	■	C	U	R	E	■	L	O	D	E
D	E	V	I	S	A	L	■	E	A	R	■	T	O	T
E	L	E	C	T	R	O	D	Y	N	A	M	I	C	S
R	O	T	■	R	E	T	I	E	■	T	A	N	K	A
S	P	Y	■	A	A	H	E	D	■	A	N	G	S	T

Puzzle Solution 18

T	R	A	M	■	U	L	C	E	R	■	U	E	Y	S
H	O	L	O	■	R	E	A	D	A	■	N	S	E	C
E	M	O	R	Y	B	O	A	R	D	■	I	T	L	L
T	A	H	O	E	■	I	L	E	A	■	V	I	P	S
A	S	A	C	A	T	■	A	D	R	E	A	M	■	■
■	■	■	C	S	I	S	■	■	E	C	A	S	H	■
B	E	S	O	■	L	A	N	A	R	K	■	T	A	E
E	D	M	■	A	T	T	I	R	E	S	■	E	L	Y
O	E	O	■	O	H	S	N	A	P	■	T	S	K	S
F	R	O	C	K	■	■	L	I	L	Y	■	■	■	■
■	■	T	W	I	T	C	H	■	N	U	M	E	R	O
F	Q	H	P	■	R	A	I	I	■	S	P	O	O	R
O	N	E	O	■	A	F	T	E	R	T	A	S	T	E
R	A	N	S	■	L	E	O	R	A	■	N	I	C	O
M	Y	S	T	■	A	S	R	E	D	■	I	N	S	S

Puzzle Solution 19

N	I	B	S		A	C	E	R	B		A	L	S	O
I	S	E	E		L	A	R	U	E		S	E	A	L
N	A	T	I	O	N	A	L	P	A	S	T	I	M	E
A	K	A		P	I	L	E			T	E	A	M	O
			B	O	C	A		T	P	E	R			
E	N	D	U	R	O		D	I	A	M	O	N	D	S
C	O	U	R	T		L	E	L	Y			F	R	E
C	A	S	T	O	R	A	N	D	P	O	L	L	U	X
Z	I	T			I	P	S	E		C	R	E	P	E
T	R	Y	A	C	A	S	E		S	T	O	R	E	R
			Q	U	A	E		G	A	R	N			
S	T	A	U	B			T	I	T	O		H	O	W
S	O	C	I	A	L	L	U	B	R	I	C	A	N	T
E	N	O	L		B	E	R	E	A		P	H	E	W
T	O	G	A		S	A	N	D	P		R	A	D	O

Puzzle Solution 20

S	E	N	D		M	I	S	E	R		O	C	T	O	P	I
A	C	A	I		A	B	A	T	E		R	H	O	T	I	C
B	U	T	T	O	N	E	D	U	P		G	O	A	T	E	E
B	A	I	Z	E		R	I	D			A	R	D	O	R	S
A	D	V	E	R	T	I	S	E	M	E	N	T	S			
T	O	E	S		H	A	M		E	N	O	L		O	V	A
H	R	S		L	A	N		B	R	U	N	E	T	T	E	S
			R	A	N		F	A	I	R			H	I	E	S
	Q	U	I	C	K	W	I	T	T	E	D	N	E	S	S	
O	U	R	S		O	B	I	S		I	O	N				
P	I	E	C	E	W	O	R	K		D	V	D		T	O	P
S	P	A		M	I	D	I		M	O	O		A	R	C	O
			L	A	R	Y	N	G	E	C	T	O	M	I	E	S
A	M	M	I	N	E		E	L	K		F	E	T	A	S	
S	E	A	M	A	T		B	R	E	A	S	T	B	O	N	E
C	A	P	I	T	A		O	M	E	G	A		A	N	I	S
I	N	S	T	E	P		A	S	S	E	T		S	E	C	S

Puzzle Solution 21

S	C	U	P	■	R	E	M	A	I	N	S	■	C	P	A
T	O	N	E	D	■	E	L	I	S	I	O	N	■	R	E	D

S C U P ■ R E M A I N S ■ C P A
T O N E D ■ E L I S I O N ■ R E D
A R I S E ■ B U R K I N A F A S O
I N S O L U B L E ■ ■ F I V E R
N E O ■ U S E ■ O S C U L A T E
■ A N O D E ■ N E W E L ■ E T A S
■ ■ P E R S I A N C A T ■ ■
D I E T ■ A P R ■ S A H A R A
D O M I N I C A N R E P U B L I C
T U S C A N ■ E A R ■ O B O E
■ ■ G U L F S T R E A M ■
A C A I ■ R O U T E ■ U M B R A
B E S T R E W N ■ P R E ■ A N I
L Y C E E ■ A A R O N S R O D
E L E M E N T A L L Y ■ T H E M E
S O N ■ V I O L A T E ■ S O F I A
T N T ■ E M P E R O R ■ E Y E S

Puzzle Solution 22

A M A H ■ P S I ■ R I S C
D I V A ■ P I P S ■ G E N O A
A N O N ■ A N A L ■ A L K Y D
R I N G F E N C E ■ G O Y A
■ A L L A Y ■ D E C
A S T R A L ■ D I S A V O W
M P H ■ R A T T A N ■ T A R E
A R O S E ■ O A T ■ G E L I D
Z I N C ■ G A T E A U ■ V E G
E G G H E A D ■ A L L E L E
■ M A D ■ M A C L E
■ S C A R ■ P A C H Y D E R M
C H I L L ■ A N T E ■ G A I A
F A T T Y ■ L I O N ■ E S P Y
C H E Z ■ E A R ■ R Y E S

Puzzle Solution 23

P	E	T	A	L	S			T	B	S	P			S	L	E	P	T
I	C	E	C	A	P			O	R	E	O			P	O	L	E	R
C	H	A	M	B	E	R	P	O	T	S				I	R	A	T	E
T	O	R	E			L	E	I				T	A	C	I	T	U	S
			S	E	L	F	C	O	M	P	L	A	C	E	N	T		
E	V	E			L	E	U			N	O	O	K			A	R	I	L
B	E	D	S	I	D	E	M	A	N	N	E	R			S	A	E	
B	L	E	A	T			L	A	V			E	N	I	D			
S	A	N	K	I	N			Y	E	T			E	B	O	O	K	S
			I	S	I	S			R	I	P			B	E	L	I	E
S	A	P			T	E	T	R	A	C	H	L	O	R	I	D	E	
A	Q	U	A			C	R	A	G			E	E	N			O	D	D
B	U	L	L	H	E	A	D	E	D	N	E	S	S					
B	A	P	T	I	S	T			R	O	W			L	O	G	E	
A	R	I	A	N			E	S	C	A	L	A	T	I	O	N	S	
T	I	E	I	N			G	O	O	K			R	A	M	P	U	P
H	A	R	R	Y			Y	U	L	E			D	R	E	S	S	Y

Puzzle Solution 24

C	H	E	F	S			D	U	K	E			E	R	O	S
G	E	N	O	A			U	R	E	A			N	E	R	O
I	N	E	R	T			A	G	E	S			D	H	A	L
			C	O	E	L	E	N	T	E	R	A	T	E		
O	T	T	E	R	S					R	U	N	E	S		
P	R	O	F	I	T	S	H	A	R	I	N	G				
A	I	R	E	S			H	E	R	O	N					
L	O	R	D			W	A	G	E	D			C	A	G	E
			T	O	K	E	N			C	O	C	O	A		
		S	C	H	O	O	L	T	E	A	C	H	E	R		
F	R	E	Y	A					O	T	H	E	R	S		
L	O	N	G	I	T	U	D	I	N	A	L					
U	P	O	N			A	G	I	N			R	E	V	U	P
M	E	R	E			C	L	O	T			R	A	I	T	A
E	R	S	T			T	Y	R	O			H	E	M	E	N

Puzzle Solution 25

W	A	R	P			O	M	B	R	E			I	R	A	D	E	S
A	S	E	A			V	A	L	O	R			C	O	P	O	U	T
L	I	P	R	E	A	D	I	N	G			E	M	I	G	R	E	
K	N	E	A	D			I	T	D			R	A	N	S	O	M	
W	I	N	D	O	W	S	H	O	P	P	I	N	G					
A	N	T	E		R	O	E		R	A	N	I			R	E	D	
Y	E	S		M	A	N		T	O	N	K	A	B	E	A	N		
	R	U	T		E	A	S	E			R	A	T	A				
	V	A	U	G	H	A	N	W	I	L	L	I	A	M	S			
B	E	N	D		V	A	S	T		E	R	E						
B	R	I	D	E	T	O	B	E		A	V	E		A	S	S		
C	B	S		V	E	I	L		O	W	E		I	N	C	A		
	N	I	L	D	E	S	P	E	R	A	N	D	U	M				
S	A	L	I	N	E		T	E	L		P	R	O	M	O			
U	N	E	S	C	O		F	I	R	E	S	T	O	R	M	S		
D	O	N	E	E	S		C	L	O	S	E		A	R	E	A		
S	A	D	I	S	T		C	E	N	S	E		D	A	D	S		

Puzzle Solution 26

S	M	O	K	E		I	D	E	S		V	E	L	A
T	I	B	I	A		N	A	S	A		I	R	A	N
A	N	E	N	T		L	I	O	N	I	Z	I	N	G
B	U	R	G	O	M	A	S	T	E	R		G	O	O
L	E	O		N	E	W		E	R	E		E	L	L
E	T	N	A		T	S	A	R		A	R	I	A	
		F	O	R		M	I	S	S	I	O	N		
	V	I	V	I	S	E	C	T	I	O	N			
	S	A	R	A	C	E	N		U	R	L			
S	I	N	E		A	D	A	R		I	O	N	S	
I	R	A		P	A	L		D	D	T		R	O	W
E	L	D		A	M	E	T	H	Y	S	T	I	N	E
R	O	A	D	M	O	V	I	E		A	H	E	A	D
R	I	T	E		L	E	E	R		R	A	N	G	E
A	N	E	W		E	L	S	E		S	I	T	E	S

Puzzle Solution 27

I	T	E	M	S		H	A	B	I	T	S		R	O	A	M
N	O	R	I	A		I	S	A	I	A	H		U	P	T	O
S	U	R	R	E	P	T	I	T	I	O	U	S	N	E	S	S
I	C	A	O		E	L	F			T	O	N	N	E	S	
T	A	T		A	W	E		B	R	O		R	E	L	A	Y
U	N	A	S	C	E	R	T	A	I	N	A	B	L	E		
			I	C	E		O	L	D	E	R		S	T	A	R
O	F	A	G	E		A	R	C		D	I	P		T	O	E
F	O	U	N	D	A	T	I	O	N	G	A	R	M	E	N	T
F	A	T		E	S	T		N	E	E		O	A	R	E	D
S	M	O	G		T	I	P	I	S		S	L	Y			
		P	R	A	I	R	I	E	S	C	H	O	O	N	E	R
I	B	S	E	N		E	N	S		R	I	G		O	R	E
M	A	Y	A	N	S			E	E	R		T	W	A	S	
P	R	I	V	A	T	E	E	N	T	E	R	P	R	I	S	E
E	R	N	E		A	R	R	A	C	K		R	E	S	E	T
L	E	G	S		G	R	A	P	H	S		O	Y	E	R	S

Puzzle Solution 28

C	A	S	S	A	T	A		S	C	H	N	A	P	S
A	P	O	L	L	O	S		C	R	E	E	P	U	P
S	P	L	I	T	U	P		R	E	A	P	P	L	Y
E	L	O	P	E	R		G	O	S	P	E	L	S	
S	E	N	O	R		D	A	D	S		T	E	A	L
		N	A	S	A	L		R	A	T	T	Y		
I	N	E	S	T	I	M	A	B	L	E		I	I	I
N	I	P		I	R	O	N	O	U	T		N	O	N
E	T	A		V	E	N	T	I	L	A	T	I	N	G
P	R	U	D	E		I	N	U	R	E				
T	A	L	A		B	A	N	G		D	A	C	E	S
	T	E	N	S	I	V	E		P	A	R	I	A	H
V	I	T	I	A	T	E		G	A	N	G	S	T	A
I	N	T	E	G	E	R		A	N	T	A	C	I	D
A	G	E	L	E	S	S		L	E	S	S	O	N	S

Puzzle Solution 29

S	H	O	P		L	A	D	L	E		G	E	T	S	O	N
N	E	R	O		E	M	E	E	R		A	S	S	U	R	E
I	T	A	L	I	A	N	A	T	E		S	C	A	R	E	S
P	A	T	I	O		E	R	G			B	A	R	E	S	T
P	E	R	C	U	S	S	I	O	N	C	A	P	S			
E	R	I	E		A	T	E		U	R	G	E		S	P	A
T	A	X		S	T	Y		N	A	U	S	E	A	T	E	D
			S	K	Y		S	A	N	S			O	Y	E	Z
	P	R	A	I	R	I	E	S	C	H	O	O	N	E	R	
T	O	I	L		R	E	T	E		L	I	E				
A	D	V	E	R	S	I	T	Y		M	I	L		A	P	P
U	S	E		W	A	S	H		H	I	V		C	R	E	E
			C	A	T	H	E	R	I	N	E	W	H	E	E	L
A	M	M	I	N	E		E	N	E		A	I	O	L	I	
G	O	U	R	D	E		B	E	D	R	A	G	G	L	E	S
O	I	L	C	A	N		U	V	E	A	L		O	A	R	S
G	L	E	A	N	S		M	E	R	L	E		E	R	S	E

Puzzle Solution 30

L	I	E	S		C	H	A	D		S	A	Y	S	O
E	D	G	E		O	O	Z	E		U	V	U	L	A
A	E	O	N		N	E	O	N		D	E	G	A	S
N	A	S	D	A	Q		T	Y	P	O		O	P	T
			E	D	U	C	E		E	K	E	S		
E	M	B	R	O	I	L		N	U	L	L	A	H	
G	O	E	S		S	A	L	P	A		M	A	T	E
R	U	T		T	W	I	R	L			V	O	X	
E	T	A	L		A	S	S	E	T		O	I	L	Y
T	H	R	E	A	D		E	Y	E	B	A	L	L	
		H	O	B	O		K	N	A	V	E			
S	T	Y		O	R	E	O		R	E	L	A	T	E
P	I	T	T	A		M	A	D	E		I	C	E	D
E	T	H	E	R		I	L	I	A		S	H	E	D
D	O	M	E	D		T	A	G	S		K	E	N	O

Puzzle Solution 31

S	L	A	V		F	E	A	T		I	M	B	U	E
N	O	N	E		U	N	D	O		N	A	I	R	A
O	U	T	L	A	N	D	E	R		L	I	N	E	R
T	R	I	A	L	R	U	N		P	A	N	D	A	S
			C	U	E		B	A	W	L				
S	A	L	M	O	N		C	I	G		Y	E	A	H
T	R	O	U	T		O	O	Z	E		L	A	Y	
R	O	O	M	T	E	M	P	E	R	A	T	U	R	E
U	M	P			R	E	S	T		L	E	D	O	N
M	A	Y	A		A	G	E		G	A	L	E	N	A
			M	E	S	A		B	A	R				
U	P	S	I	D	E		M	A	R	M	I	T	E	S
S	O	N	D	E		D	U	S	T	S	T	O	R	M
P	R	I	S	M		A	L	E	E		C	L	I	O
S	E	P	T	A		H	E	R	R		H	U	N	G

Puzzle Solution 32

P	F	F	T		D	O	D	O		I	S	B	A	D
E	I	L	A		E	T	C	S		N	O	A	I	R
A	S	O	B		W	A	C	K		F	U	D	D	Y
T	H	E	L	A	D	Y	V	A	N	I	S	H	E	S
			A	R	R		R	A	N					
A	G	R		U	O	F	A		W	I	R	E	R	S
C	O	H	O		P	O	L	E		T	O	R	I	I
I	D	Y	L	L	S	O	F	T	H	E	K	I	N	G
N	E	M	E	A		T	I	R	O		S	E	T	I
G	L	E	N	N	M		O	E	N	S		S	T	L
			T	S	A		E	N	C					
T	R	Y	B	E	F	O	R	E	Y	O	U	B	U	Y
R	E	A	I	R		L	I	S	P		R	O	T	A
A	L	L	E	N		E	L	O	I		B	O	N	Y
S	O	U	R	S		R	E	S	E		S	E	E	A

Puzzle Solution 33

H	A	C	K		W	O	A	D			G	I	N	A
A	G	O	N		A	R	R	E	A		O	F	A	N
D	U	P	E	T	H	E	T	R	U	S	T	I	N	G
J	A	Y		I	O	L	E		S	C	H	M	O	E
	S	K	O		E	S	A	I						
B	R	I	T	I	S	H	R	A	I	N	C	O	A	T
R	O	C	K	S		O	C	T	E	T		N	T	H
E	L	I	S		S	U	P	E	S		K	I	R	I
A	L	L		I	N	S	T	R		L	I	C	E	S
K	A	Y	S	E	Y	E	S	S	E	E	D	E	E	S
	L	A	D	D		R	T	S						
I	L	D	U	C	E		P	S	E	C		A	P	B
P	R	E	S	H	R	U	N	K	S	H	I	R	T	S
S	O	P	H		S	T	O	A	S		E	N	A	M
E	N	T	Y		A	M	T	O		L	O	S	T	

Puzzle Solution 34

O	R	C	H		E	L	E	N	A		H	A	A	G
P	E	L	E		T	A	R	O	S		E	R	N	A
S	C	A	B		H	D	T	V	S		A	M	O	S
	U	R	B	A	N	D	E	O	D	O	R	A	N	T
M	L	I		R	O	S		V	A	N	Y	A		
A	D	N	A	N		N	A	B	E		I	M	P	
L	E	E	K		S	C	A	L	A	R	S			
A	R	T	I	F	I	C	I	A	L	T	O	O	T	H
	N	I	R	V	A	N	A		A	C	H	E		
M	M	L		R	E	I	D		F	R	O	E	S	
Y	E	A	S	T		G	M	S		T	A	S		
F	E	T	C	H	I	N	G	O	U	T	F	I	T	
O	T	R	A		L	E	A	D	S		D	L	E	G
O	E	I	L		L	I	E	B	E		I	L	R	E
T	R	A	Y		E	N	D	E	D		C	O	S	T

Puzzle Solution 35

```
D A H L   M O N T T   J A V A
A L I A   A D O R A   A D A R
L O L L   I D I O M   C A P A
  S T A I N E D U P C O M I C
  L E E R     A U B A D E
L A M A R R   R A B B I
R E I N E   T A T A S   E Y E
O R A D   S A T A Y   E B E N
N I S   S A H I B   I A L S O
  N E U R O   A S S A I L
I U O E P C   D D A Y
V K P S A I Z K R D T A W H
E I N S   E A N O L   C A T O
R A I I   S C A P E   E K E D
S H N E   T H R A D   S E N D
```

Puzzle Solution 36

```
A T H S   F E L I   S C H W A
P E A K   A T E N   A A M P G
A X C E   K A N S   U B O A T
L A K E T I T I C A C A
  S S T A R S   N E L S O N
  E B S   M O A   A C T O
S N A R L   D E N I S   O E D
N O R S E L I T E R A T U R E
A N I   T E R R I   T E R I S
P O O L   O T O   N R A
E S T H E R   B E A T E N
  A N A T O L E P A R I S
L E G S D   A B U D   B R O W
P A L A U   L I T E   L O B E
S T O N E   L E H R   E L E E
```

Puzzle Solution 37

T	U	T	O	■	R	A	E	S	■	L	A	M	P	B
A	S	I	N	■	E	N	D	O	■	O	D	E	A	R
M	O	N	T	■	T	O	D	D	■	V	I	T	A	E
I	F	Y	O	U	S	A	Y	O	N	E	M	O	R	E
L	A	T	E	N	Y	■	I	M	A	Y	■	■	■	
■	■	A	N	I	N	■	R	A	N	S	O	M		
A	N	S	E	R	■	E	G	G	Y	■	A	R	E	O
G	L	I	M	M	E	R	■	O	A	U	C	T	I	O
E	R	T	E	■	N	I	S	I	■	S	H	A	L	T
S	B	A	R	R	O	■	E	N	C	E	■	■	■	
■	■	U	R	G	E	■	R	I	N	S	E	R		
P	E	O	P	L	E	S	P	R	I	N	C	E	S	S
A	R	O	N	I	■	T	A	H	E	■	A	L	T	I
L	E	N	I	N	■	A	G	E	R	■	A	M	A	D
L	I	A	N	G	■	R	E	O	S	■	S	A	B	E

Puzzle Solution 38

T	I	M	E	L	A	G	■	E	D	D	A	■	A	D	O	S
O	C	A	R	I	N	A	■	N	O	I	R	■	V	I	V	E
P	O	T	A	T	O	B	E	E	T	L	E	■	A	S	E	A
E	N	S	■	T	I	L	E	■	T	E	N	■	T	A	R	
■	■	G	E	N	E	R	A	L	M	A	N	A	G	E	R	
I	M	P	A	R	T	■	G	E	M	■	A	R	R	A	Y	
D	I	R	T	■	S	O	A	R	■	A	F	T	■	E	T	A
S	L	O	E	S	■	O	R	E	O	■	D	U	R	E	S	S
■	■	C	A	T	C	H	M	E	N	T	A	R	E	A		
M	A	T	U	R	E	■	S	A	T	E	■	E	L	B	O	W
A	G	O	■	O	P	S	■	B	O	A	R	■	A	L	A	R
R	I	L	E	D	■	T	E	L	■	E	A	T	E	R	Y	
S	T	O	R	E	D	E	T	E	C	T	I	V	E	■	■	
■	A	G	E	■	O	W	E	■	U	R	S	A	■	A	C	E
O	T	I	C	■	P	A	R	A	D	I	S	I	A	C	A	L
C	O	S	T	■	E	R	N	S	■	P	U	L	S	A	R	S
A	R	T	S	■	S	T	E	P	■	S	E	S	S	I	L	E

Puzzle Solution 39

A	L	B	S	■	A	L	T	O	S	■	L	I	S	P
S	A	R	I	■	L	E	A	S	T	■	A	N	T	E
I	D	O	L	■	B	I	B	L	E	■	T	S	A	R
F	Y	K	E	S	■	L	O	N	G	T	E	R	M	■
■	E	X	C	U	S	E	■	C	H	I	C	■	■	■
P	U	N	■	A	R	C	O	■	H	E	C	T	O	R
A	S	H	■	D	E	A	F	■	E	E	R	I	E	■
G	A	E	L	■	A	R	C	E	D	■	S	E	L	F
A	G	A	I	N	■	O	V	E	R	■	P	E	E	■
N	E	R	V	E	D	■	N	E	V	E	■	E	R	R
■	■	T	E	A	R	■	T	R	A	V	E	L	■	■
A	P	E	R	T	U	R	E	■	S	A	L	A	L	■
M	I	D	I	■	D	Y	N	E	S	■	V	E	D	A
I	S	L	E	■	G	E	T	T	O	■	E	N	D	S
D	A	Y	S	■	E	S	S	A	Y	■	S	T	Y	E

Puzzle Solution 40

T	A	R	P	■	P	S	S	T	■	S	N	O	B	■
O	D	O	R	■	I	C	H	O	R	■	H	I	R	E
D	A	T	E	■	N	A	I	R	A	■	U	S	E	R
D	R	A	C	O	■	G	R	A	D	A	T	I	O	N
■	■	R	E	N	D	■	R	H	I	N	O	■	■	■
S	T	Y	P	T	I	C	■	S	I	T	U	P	S	■
P	E	C	T	O	R	A	L	■	S	T	R	O	P	■
U	R	L	■	T	R	I	T	E	■	O	N	O	■	■
D	R	U	I	D	■	P	H	Y	S	I	C	A	L	■
■	A	B	R	O	A	D	■	Y	A	M	M	E	R	S
■	A	T	S	E	A	■	S	U	P	S	■	■	■	■
D	E	A	T	H	T	O	L	L	■	T	A	S	T	E
A	C	R	E	■	I	D	I	O	T	■	S	I	O	N
S	H	U	L	■	R	A	B	B	I	■	T	O	L	D
H	O	M	Y	■	R	I	S	C	■	O	N	U	S	■

Puzzle Solution 41

E	L	O	P	E	R		C	G	I		A	M	P	S
R	E	P	L	A	Y		U	R	N		B	E	A	U
N	A	T	U	R	A	L	L	A	N	G	U	A	G	E
E	R	S	T			I	M	F		O	T	T	E	R
			O	B	O	E		F	U	R				
M	P	H		R	E	N	A	I	S	S	A	N	C	E
O	R	A	T	O	R		S	T	E	E	P	E	R	S
T	O	U	R			S	K	I			E	V	E	S
O	B	S	O	L	E	T	E		T	E	R	E	T	E
R	E	A	D	Y	T	O	W	E	A	R		R	E	X
			C	A	R		M	I	R	V				
A	L	G	A	E		A	N	I			E	R	G	O
M	A	R	K	E	T	G	A	R	D	E	N	I	N	G
A	C	A	I		E	E	R		A	V	A	T	A	R
H	Y	M	N		A	S	K		D	E	L	E	T	E

Puzzle Solution 42

A	B	L	E		A	P	P	E	A	R		S	A	C
G	O	E	R		C	U	R	A	R	E		T	I	E
A	N	A	S	T	O	M	O	S	I	S		U	R	L
M	I	D	T	E	R	M		E	A	T	A	B	L	E
A	T	E		A	N	E	W		I	B	I	S		
S	O	R	E	L		L	E	S	T		L	I	F	T
			R	E	S		T	H	O	U		E	T	A
	B	R	A	I	N	T	E	A	S	E	R			
A	C	E		F	L	U	E		D	U	B			
L	A	M	P		T	B	S	P		A	B	A	C	A
T	R	I	O			T	A	E	L		M	O	M	
H	A	R	I	S	S	A		S	A	L	A	B	L	E
O	B	I		C	O	U	N	T	R	Y	C	L	U	B
R	A	N		O	S	T	E	A	L		M	E	G	A
N	O	G		T	O	O	T	S	Y		E	R	O	S

Puzzle Solution 43

A	L	E	P	H		H	A	S	P		P	U	F	F
R	O	M	E	O		E	G	I	S		O	G	E	E
C	O	U	R	T	M	A	R	T	I	A	L	L	E	D
O	P	S		D	E	V	A		V	E	I	L	S	
		A	O	N	E		P	O	E	M				
S	H	O	T	G	U	N	M	A	R	R	I	A	G	E
N	A	V	E	S		A	N	A		C	H	A	R	
A	L	A		R	A	T	E	L		O	V	A		
R	A	R	E		A	R	C		C	O	L	E	S	
F	L	Y	O	F	F	T	H	E	H	A	N	D	L	E
	L	O	T	S		N	U	D	E					
S	T	A	I	R		D	E	L	E		T	A	U	
C	O	N	T	E	M	P	O	R	A	N	E	O	U	S
A	M	A	H		P	A	N	G		Z	A	I	R	E
G	E	L	S		G	R	A	Y		A	T	L	A	S

Puzzle Solution 44

S	C	A	G		A	S	H		T	I	A	R	A	
N	A	S	A		S	O	O	K		A	R	M	O	R
A	N	T	E		A	F	A	R		X	E	B	E	C
F	O	O	L	S	P	A	R	A	D	I	S	E		
U	N	R	I	P		D	U	E		R	N	A		
	C	A	R	P	E	T	B	A	G	G	E	R		
R	O	M		A	I	D		A	R	A	R	A	T	
A	M	A	T	I	V	E		S	U	M	M	I	T	S
M	A	S	A	L	A		D	O	C		S	H	Y	
P	H	O	T	O	G	R	A	P	H	E	R			
S	A	C		E	E	N		N	E	E	D	S		
	H	O	R	S	E	C	H	E	S	T	N	U	T	
A	L	I	B	I		D	I	O	R		I	D	L	E
A	E	S	O	P		S	N	O	G		R	O	S	E
H	A	M	E	S		G	P	O		E	W	E	R	

Puzzle Solution 45

```
C H O M P     T R A S H
W O O D I E R   S H O T P U T
A R G O N N E   P E D A L E R
K Y A N I T E   H O S T I L E
E Z R A   V E E S   I C E S
N A T T Y   E U R O   M E S S
  S H A L L   R I P P E R S
        E A R A C H E
  G R A M M E S   Y E M E N
O R A L   I B I S   R E V U P
C A G E   N A A N   M E R E
E N T R E A T   I M M E N S E
A G I T A T E   P O I N T E R
N E M E S E S   E A S T E R S
  R E D E D     S T O R Y
```

Puzzle Solution 46

```
A R M   M A S T I C S   S A R A H
R U E   O P H I D I A   E R O D E
A B S E N T E E I S M   R O O D S
B E A U     D O C   S O M M E
      R O C K Y M O U N T A I N S
C A R O T E N E   S A Y   E D O
A C E   T I E   P U E R P E R A L
B A N D O L E E R S   L E T
S I T E   L A Y E R   C O R A
    L A V   S E D A N C H A I R
A R T I F I C E R   D I E   R P M
K O I   L O W   W A L R U S E S
A S S A U L T C O U R S E S
  S A N T A   U P S   E A S T
P I N O T   D R E S S C I R C L E
A N E L E   D I N E O U T   M A D
R I S E R   T O S S U P S   E M S
```

Puzzle Solution 47

S	N	O	W		P	L	E	B		A	I	S	L	E
T	I	R	O		H	O	M	E		L	I	P	I	D
E	S	A	U		A	B	E	T		P	I	E	C	E
P	I	L	L	A	R		R	A	J	A		E	E	N
		D	I	M	L	Y		I	C	E	D			
S	C	A	N	D	A	L		G	A	R	L	I	C	
C	O	P	T		C	A	L	L	S		A	I	N	U
A	M	P		O	M	A	H	A			M	A	R	
B	E	E	P		P	A	P	A	W		P	I	N	E
S	T	A	T	U	E			S	P	O	R	T	E	D
	S	A	R	I		M	A	U	V	E				
A	H	A		G	A	Z	A		Z	A	F	T	I	G
S	A	B	L	E		O	Y	E	Z		A	B	L	E
A	L	L	E	N		O	V	A	L		C	A	L	L
P	L	E	A	T		M	E	R	E		E	R	S	T

Puzzle Solution 48

A	L	B	S		R	E	S	T	S		S	H	R	O	V	E
B	A	R	T		E	X	T	R	A		H	E	A	L	E	R
S	C	O	R	E	S	H	E	E	T		R	A	G	L	A	N
C	U	M	I	N		A	L	A		I	D	E	A	L	S	
I	N	A	D	E	Q	U	A	T	E	N	E	S	S			
S	A	T	E		U	S	E		D	Y	K	E		O	V	A
E	R	E		B	A	T		O	I	L	S	T	O	N	E	S
		C	O	S		A	L	T	O			M	U	S	S	
	M	A	R	X	I	S	T	L	E	N	I	N	I	S	T	
G	A	I	A		L	O	A	D		N	U	T				
M	Y	R	M	I	D	O	N	S		P	U	B		S	A	C
T	O	Y		N	A	P	A		M	A	R		S	I	L	L
	I	N	T	E	L	L	I	G	E	N	T	S	I	A		
H	E	R	N	I	A		I	C	E		O	A	T	E	N	
O	C	E	A	N	S		S	T	R	A	G	G	L	I	N	G
O	R	A	N	G	E		P	H	O	N	S		K	N	E	E
P	U	R	E	S	T		Y	E	N	T	A		S	E	E	D

Puzzle Solution 49

T	B	S	P		A	C	A	I		A	V	A	I	L
O	L	L	A		N	O	R	M		S	I	N	C	E
R	O	U	N	D	T	R	I	P		P	E	N	E	S
T	A	S	T	E		P	L	E	B	E		E	S	T
E	T	H		C	P	U		T	E	N	S			
			S	E	I	S	M	I	C		H	E	R	B
E	M	E	T	I	C		A	G	O		O	A	H	U
D	I	G	I	T	A	L	C	O	M	P	U	T	E	R
G	R	I	N		Y	A	H		I	L	L	S	A	Y
E	A	S	T		U	N	O	W	N	E	D			
			S	E	N	D		A	G	A		R	O	C
R	N	A		T	E	M	P	T		S	T	A	R	R
S	O	U	G	H		A	R	T	L	E	S	S	L	Y
V	I	R	E	O		S	I	L	O		A	T	O	P
P	L	A	T	S		S	M	E	W		R	A	N	T

Puzzle Solution 50

L	A	T	T	E		D	R	E	A	D		G	S	A
A	C	O	R	N		E	E	R	I	E		R	A	N
P	H	R	A	S	A	L	V	E	R	B		A	R	T
P	E	E	N		N	E	E		R	U	P	E	E	
			C	U	T	T	L	E	F	I	S	H	E	S
W	H	E	E	L	I	E		A	R	E	A			
H	A	D	S	T		G	R	A	F	F	I	T	I	
I	T	D		G	U	E	S	T		D	E	C		
R	E	A	D	J	U	S	T		C	R	O	A	K	
			Y	U	L	E		U	S	U	A	L	L	Y
S	N	E	A	K	P	R	E	V	I	E	W			
H	O	R	D	E		N	U	T		H	Y	P	O	
U	R	N		B	A	T	T	L	E	F	I	E	L	D
S	I	S		O	P	E	R	A		E	D	G	E	D
H	A	T		X	R	A	Y	S		D	E	G	A	S